THE MANAGEMENT OF INNOVATION

The Management of Innovation

TOM BURNS
and
G. M. STALKER

OXFORD UNIVERSITY PRESS
1994

Oxford University Press, Walton Street, Oxford OX2 6DP

Oxford New York
Athens Auckland Bangkok Bombay
Calcutta Cape Town Dar es Salaam Delhi
Florence Hong Kong Istanbul Karachi
Kuala Lumpur Madras Madrid Melbourne
Mexico City Nairobi Paris Singapore
Taipei Tokyo Toronto

and associated companies in
Berlin Ibadan

Oxford is a trade mark of Oxford University Press

Published in the United States
by Oxford University Press Inc., New York

© Tom Burns and G. M. Stalker 1961
Revised edition published 1994

British Library Cataloguing in Publication Data
Data available

Library of Congress Cataloging in Publication Data
Data available
ISBN 0-19-828877-8
ISBN 0-19-828878-6 (Pbk)

Printed in Great Britain
on acid-free paper by
Biddles Ltd.,
Guildford and King's Lynn

CONTENTS

PREFACE TO THE THIRD EDITION

The Management of Innovation was first published in 1961. The distance in time is not all that great, but we are perhaps far enough off to recognize the period, just short of thirty years, from 1945 to the early 1970s, as quite different in character not only from the inter-war years but also from the two subsequent decades. At the very least, the political, economic, and social circumstances seem, in retrospect, altogether different from now. By consequence—and it is this that makes the really significant difference—the horizon of possibilities and expectations in terms of social change and improved economic welfare was much wider than it has become since then.

The years when the research was done and the book written and published came just midway through that post-war period. The first two prefaces gave some account of the circumstances which prompted the research, but both of them are almost as old as the book itself. This being so, it has occurred to me that this third preface might best be devoted, first of all, to some brief sketch of 'how it was', in a wider context than that adopted for the earlier prefaces, and of the kind of direction taken by the study of organizations in general during the 1950s and 1960s.

I

The Management of Innovation was never a best-seller, even by the lowly standards of 'academic' books, but it did go on to have a fairly long shelf-life, outlasting the lifetime of the book on the organization of the BBC which was meant as the next step.

It was followed in fairly close order by published accounts of research by Joan Woodward in the south of England and by Michel Crozier in France. There were striking similarities about the conclusions of all three of them, but it was obvious that they had been arrived at independently; none of us knew anything of the existence of the others, let alone of what they were up to.

This, at first sight, may seem a little odd. It is of course commonplace enough in the world of natural science for discoveries, break-throughs, new syntheses, of major importance to occur almost simultaneously; indeed, fear of its happening can become a major preoccupation (sometimes a perpetual nightmare) for the laboratory scientist. The world of the social sciences, so far as its academic and research sector is concerned, is nothing like this. The three books, Woodward's *Industrial Organization: Theory and Practice*, Crozier's *The Bureaucratic Phenomenon*, and my own, were very different in terms of subject-matter, scope, and detailed findings. We were simply not aware of what the others were up to. What similarity there is resides in the critical approach we all took towards generally accepted notions of the nature, structure, and processes of industrial organization. And this came, first of all, simply from our common experience, along with all our contemporaries, of the economic, political, and social conditions prevailing in Europe (and America).

The end of the 1950s marked the culmination of a period which began with the massive task of post-war physical reconstruction and political renovation and moved on into economic expansion and technological innovation on an equally massive scale. This was true not only of industrial and commercial activity but also socially and culturally; television and air-travel seemed to have played no less a part in transforming the character of the world than electronics and nuclear power or the 'economic miracles' in Germany and Japan and the equally miraculous experience of full employment elsewhere.

Economic and political circumstances can also be seen, in retrospect, as underlying the prominent part played by the idea of social cohesion and equilibrium in sociology and economics during the first post-war years— or, rather, the widespread concern to develop conceptions and theories about the social and economic system which would help preserve them. On the one side, there was the rapid progress achieved in reconversion and reconstruction as compared with any previous post-war period; the commencement of a period of prosperity unparalleled in the past and in especial contrast with the pre-war years; the boom in house building and babies; the explosion in the quantity and variety of consumer goods and services which transformed domestic life—all these seemed to underwrite the value of stability, commitment to the status quo, and conformity. On the other hand, and at the same time, the bomb and the cold war revived all those fears of mutual destruction between societies and self-destruction by society which had pervaded the 1930s and the war years.

The whole circumstantial postulate and the lesson to be drawn from it were summarized by Louis Wirth in his presidential address to the American Sociological Society in 1947: 'The subject matter for the cultivation of which society sustains us . . . is the life of man in society, and the heart of that subject matter is the understanding of the processes through which consensus on a world scale is created.' It was a restatement, more concrete and more confident, of the selfsame thesis to which his Chicago colleagues, Park and Burgess, had committed themselves in an introductory textbook published over twenty years earlier: 'Society viewed abstractly is an organization of individuals; considered concretely, it is a complex of organised habits, sentiments, and social attitudes—in short, consensus.'

Such an approach not only prepared the ground for the dominance of functionalist theory proper when it became fully fashioned in the post-war years but set the pattern for all the social sciences in the 1950s. Political science, for example, armed with the skills and the data of survey research, took as its 'prime concern' the stability of a specific institutional structure or political regime—'the *social* conditions of democracy'. Studies of industrial organizations in the first ten or so years after the war reveal the same devotion to what was seen as the central task of sociology and the other 'behavioural sciences'. The institutional framework of contemporary business and industry had to be accepted as given. In any case, and despite appearances to the contrary, the activities and purposes of economic organizations did further human welfare; they were an effective means of providing both the means of livelihood for the majority of people and the goods and services they wanted. Social science could best play its part by identifying the obstacles and the sources of friction which impeded the proper functioning of organizations. It might, perhaps, even go on to discover or suggest ways of removing such impediments, improve organizational efficiency and effectiveness, and so promote the public good—to which the well-being of business organizations was essential.

Roethlisberger and Dickson's *Management and the Worker* is an impressive and lasting memorial to this kind of thinking. It consummated the work of Elton Mayo and others at the Harvard Business School with twelve years of research at the Western Electric Company's plant at Hawthorne, Chicago, into 'the intangible factors in the work situation that affect the morale and productive efficiency of shopworkers', in the words of the senior executive of Western Electric who wrote the foreword to the book.

By 1950 the book had achieved the status of a minor classic.

Throughout the 1950s, the notion of the 'informal organization' (i.e. the relationships and interactions between members of work groups) which was central to the Hawthorne studies served as a receptacle for observations about the behaviour, the relationships, the sentiments and beliefs, the commitments and self-images of factory workers which were seen as irrelevant to the 'formal organization' (i.e. management) and its purposes. The dualism—almost the dichotomy—of formal and informal attracted the notions of rational and irrational as if they were cognate terms. Together they engendered a view of the 'informal organization' as pathological, 'dysfunctional', comprehending all that had defeated Taylorite attempts at human engineering. In much of the literature of industrial studies it appeared as a kind of Freudian unconscious, which it was for research to explore, bring its hidden activities and commitments and conflicts into the light of consciousness (i.e. to the notice of management), and thereby enable its misdirected efforts to be checked or harnessed to ends consonant with those of the enterprise.

The same strategy acquired a broader base after the end of World War II. Since community-mindedness, co-operation, participation in collective action, and the like were now taken to be systemically normal for normal human beings, the task of the social scientist was to identify the obstacles to effective performance in an industrial organization, and show how they might be removed.

During the 1950s, though, the balance began to shift—not immoderately, but enough to focus rather more attention on possible impediments to the efficiency and effectiveness of management and organizational structure. It was built around the criticism of the nineteenth-century notion of organizations as instruments whereby entrepreneurs of all kinds—social, political, and cultural, as well as economic, and in groups as well as individually—attained their ends by designing rational structures of decision-making and authoritative control.

The main target of this criticism was, to begin with, 'scientific management', which had swept the board from the 1890s onwards, and had captivated such disparate minds as those of Marshall, Lenin, and Weber. F. W. Taylor, the man whose name is most closely associated with this conception, had begun as an engineer, and scientific management consisted precisely of planning, directing, and monitoring factory workers in the way engineers planned, controlled, and monitored the operations and output of machines.

But Weber's own account of bureaucracy also came under criticism, although in somewhat oblique fashion. While political science was the main beneficiary of the rather fragmentary translations of Weber's writings which began to appear in the 1940s, their availability also imparted further impetus, by way of an intellectually respectable additive, to the study of industrial organization after the war.

In fact, from the early 1950s on, 'bureaucracy' came to be regarded as synonymous with the 'management structure' and the 'formal organization' of earlier years. The influence of Weber's writings is clear enough in the work of Selznick, Blau, and Gouldner, but the influence was by no means direct. Much of the research which was certainly stimulated by, and perhaps originated in, Weber's account of authority types and social action seems to have been undertaken, or at least concluded, as a challenge to Weber's ideas rather than an application or development of them.

For the notion that people spend their working lives ruled by those set above them by virtue of the inherent rationality, as well as the law-like authority, vested in the hierarchic command structure of bureaucratic organization proved to be just as challengeable as the rational control of working behaviour imposed by scientific management. So we find organizations in which authority is *really* vested in political caucuses made up of local interest-groups, as in Selznick's *TVA and the Grass Roots*; in which quite different varieties of bureaucracy—'punishment-centred', 'mock', and 'representative'—can take turn about with each other in the management structure of a gypsum mining company, as in Gouldner's *Patterns of Industrial Bureaucracy*; or where one welfare agency can exist as a rule-bound bureaucracy alongside another founded on participation on the basis of commitment to its social role in two neighbouring city districts, as in Blau's *Dynamics of Bureaucracy*.

A similar reaction, this time against the spelling-out of the doom of oligarchy to which all democratic institutions, organizations, and movements were condemned, according to Weber's friend Michels, prompted *Union Democracy*, aimed at showing, in at least one instance, how the 'iron law of oligarchy' did not necessarily apply.

But it is important to grasp the fact that the critique of bureaucratic rationality evident in these studies also contributed to the dominant image of society as a system which functioned in terms of an equilibrium, a steady state. Even though there was manifest conflict and competition between individuals and groups within the same system, the consequence was not necessarily disruption or breakdown; it might be the prelude to structural change and, through it, a new, higher-level, equilibrium. By

1960 the favoured interpretation of conflict had reduced it to an element of social process. Conflict was properly to be envisaged as something that invigorated social institutions by promoting adaptive changes, and so raised them to higher levels of effectiveness. What was it, after all, but a more vigorous form of competition?

The same truth—in rather more moderate terms—applied to organizations. Nevertheless, intrusive or frictional elements might at times cause the equilibrium state to maintain itself at a lower level; it might be that there were ways in which the commitment of workers to the tasks management set them might be sustained or enlarged. This meant that there was therefore an understandable and acceptable role for social scientists who claimed to be in pursuit of information and experience which could be used to remedy that kind of situation.

II

By the late 1950s, the pursuit of consensus which Louis Wirth had suggested as the ultimate goal of the social sciences, was slackening. To put it at its most banal, the time was ripe for change. More to the point, the critical role of the social sciences* was reasserting itself. For what was beginning to happen within the restricted field of organization studies was part of a much wider movement in the social sciences generally, which in turn reflected the consummation of the period of post-war reconstruction and recovery. 'Labelling theory' turned the tables on established ideas of criminality and delinquency; urban sociology took on a new lease of life in Britain, France, and America; Goffman began his assault on established ideas about the self and the nature of social relationships; neo-Marxist theory took an unexpected turn towards structuralism.

The first sign of a radically new departure in organization studies came towards the end of the 1950s. J. G. March and Herbert Simon's *Organiza-*

---

* 'The purpose of sociology is to achieve an understanding of social behaviour and social institutions which is different from that current among the people through whose conduct the institutions exist; and understanding which is not merely different but new and better. The practice of sociology is criticism. It exists to criticise claims about the value of achievement and to question assumptions about the meaning of conduct. It is the business of sociologists to conduct a critical debate with the public about its equipment of social institutions.' (T. Burns, 'Sociological Explanation', inaugural lecture, Univ. of Edinburgh, 1964.)

tions turned aside from concern with organizational structure or managerial control and focused instead on what its authors saw as important features of problem-solving and of rational choice—or rather, as they put it, the characteristics of human problem-solving and rational choice as fundamental to the structure and functioning of organizations. The importance of the book, and of Simon's other contributions to organizational analysis, lay principally in the concepts of limited aspirations and 'bounded rationality'.

Motivation, individual and collective, is a matter of the balance between satisfaction and expectation. When satisfaction falls below previous expectation, or expectation rises above the previous level of satisfaction, this generates a search for ways of improving performance so that satisfaction will again match expectation. The results that people—individual members of an organization as well as the management as a whole, or groups, or departments—seek to achieve are not the best possible but those which will prove good enough to meet expectations. 'Maximizing', conventionally the guiding principle of economic (sc. managerial) man, is replaced by 'satisficing'.

There are, secondly, limits to the capacity of human intelligence to match the complexities of the problems that individuals and organizations face. So it is that, in practice, managerial decision-making ordinarily calls for simplified models that capture the main, or most urgent, features of a problem without all its complexities.

In reality, it was argued, most organizations, most of the time, run on repertories of decision-making and choice, each of which deals with the alternatives presented in constantly recurring situations. Only when these repertories of 'programmed' decision-making fail to meet the situation satisfactorily, or when people's aspirations are raised, do they actively search for better alternatives. It is by searching that new possibilities for action are discovered, and searching means setting wider boundaries for problem-solving and confronting new arrays of rational choices.

Organizations appeared in 1958, when I had almost finished writing *The Management of Innovation*. Simon's ideas derived from his training in behavioural psychology, which he had taken up after beginning his research career in the study of administrative systems in local government. My own preoccupation was with the structure and dynamics of interpersonal relationships, of the various interests pursued by individuals and of the alliances they formed to further them and the social sub-systems observably present in organizations. There was nevertheless an obvious parallel between the 'mechanistic' and 'organic' systems of management featured in the book I was working on and Simon's 'programmed' and 'non-programmed' decision-making, and the contrast he made between those circumstances in which expectations can be met by decision-making of the routine kind and those which involve novel situations, new aspirations, and raised expectations.

There was also a parallel between the ideas developed in *The Management of Innovation* and Joan Woodward's survey of firms in south Essex. She found much the same difference in management structure and relationships as those I had labelled 'mechanistic' and 'organic' in those firms engaged in large-batch production (which required fairly stable, almost routine, decision-making), and those involved in small-scale job-bing work, technical innovation, and the like, in which authoritative instruction tended to be tempered by consultation and the hierarchic order of rank much less obvious.

Organizations, The Management of Innovation, and Woodward's *Industrial Organization: Theory and Practice* have enough in common to represent one line of development out of the critique of bureaucratic organization. All three attached central importance to the existence of alternative systems of management, one appropriate to relatively stable technological and market conditions, the other to situations in which technology and market situation were changing fairly rapidly.

There was another line of development, which fastened on the prevalence of competition and conflict *within* organizations. When we come to Michel Crozier, we find criticism of conventional management structure (sc. bureaucracy) to the point of denying it the validity it had possessed as the chosen instrument of rationally designed organization. His empirical studies of two different organizations, both operated by the French government—the central Paris office of the national savings-bank ('giro') system, and the twenty-odd factories operated by the government under its monopoly of tobacco manufacture—showed up much the same pattern of perpetually recurring conflict over status, demarcation rules, operating regulations, and the like. In effect, 'bureaucracy' turns out to be the product of a vicious circle. Individual complaints were not dealt with satisfactorily, which led to the formation of dissentient groups and so to political conflict and eventually to bargaining and negotiation. It comes full circle with a revised set of old rules or a set of new ones—which, in time, provide the battleground for a new cycle of complaint, political conflict, bargaining, and settlement. In a sense, Crozier stood Michels's 'bureaucratization of politics' thesis on its head; bureaucracy was not only the outcome but also the begetter of political conflict. At least this was true of the organizations he studied; Crozier does qualify his findings to the extent of accepting that the 'vicious circle of bureaucracy' pattern owed something to a peculiarly French predilection for political confrontation, as well, of course, to both organizations being particularly susceptible to the traditional preoccupation of French administration with elaborate, rule-governed, procedures.

During the early 1960s two major studies carried the idea of competition and conflict well beyond the point reached by Selznick, Gouldner, or Crozier, representing them as central elements in the actual practice of contemporary organizations. The 'negotiated order' model of organization as first set out in *Psychiatric Ideologies and Institutions*, the report of a study of two large psychiatric hospitals in Chicago by a research team, the senior figure being Anselm Strauss. Psychiatry being what it is, the clinical specialists in charge of different departments tended to belong to different schools of thought about the appropriate management and care of patients with various forms of mental illness—hence the 'Psychiatric Ideologies' of the title. This meant that problems arose about the exercise of overall administrative control.

This was also the central idea of R. M. Cyert and J. G. March's *Behavioral Theory of the Firm*, published in 1963. In any organization, they argued, there can be, and usually are, quite different and virtually incompatible views about what the goals of the organization are. There are also— rather more frequently, and clearly—different ideas about how to attain those goals. What agreement there is tends to settle for 'highly ambiguous' goals. And behind any agreement reached on rather vague objectives, 'there is considerable disagreement and uncertainty about sub-goals'; in fact, 'organizations appear to be pursuing different goals at the same time'.

The Management of Innovation had a foot in both camps. Like Simon and Woodward, it predicated alternative systems of management, each appropriate to different rates of change in technology and market. Like Crozier, it recognized the presence of disputes about regulations, and, like Strauss, and Cyert and March, conflicting ideas about overall policy and about the appropriate strategies for meeting new situations. To these it added competition among individuals for advancement and the recurrent problem of who was to succeed the chief executive.

There was also one major difference from all of them. Woodward, for example, seemed to infer that the correspondence between the 'mechanistic' kind of system and large-batch (or 'mass') production and between an 'organic' system and small-scale or innovative production processes was more or less 'natural'. In Simon's work, too, there was the same suggestion that 'programmed' would naturally give way to 'non-programmed' decision-making when circumstances introduced novel and unfamiliar circumstances into the situation confronting managers. The discrepancy between expectations and satisfactions which this induced would also, it seemed, lead naturally to the search for fresh work arrangements, alternative products, new technology.

In my own studies, on the other hand, it was at this very point that management problems emerged. The research reported in *The Management of Innovation* began as an attempt to monitor the progress of a scheme inaugurated in the mid-1950s by The Scottish Council (Development and Industry) to facilitate the entry of established engineering firms in Scotland into the new and promising field of electronically controlled machinery and equipment. In most of the firms studied, their reluctance to depart from the management structure and operational practices to which they had become habituated proved too strong. The solutions they in practice sought to apply were mainly to reinforce the old 'mechanistic' system and—very often—to shield it from the impact of the new ventures by keeping them in separate buildings, even separate localities, as well as organizationally segregated in terms of the structure of management. The successful few had moved, or were moving, from one system to the other, seeing the need for change and rearranging the management system in pragmatic but quite effective ways.

The main theme of *Psychiatric Ideologies and Institutions* was that neither the general policy to be pursued nor overall administrative control was handled by—or through—any bureaucratic or managerial structure of authority but worked out and agreed in *ad hoc* fashion through an unending process of discussion, bargaining, and negotiation. Cyert and March make the same point; firms are constantly trying to resolve the differences of opinion that arise within management, so that instead of a clearly defined and consistently held objective, firms are involved in a constant process of internal debate and negotiation. In practice, the goals of any organization are being constantly amended as the result of what they call a 'continuous bargaining-learning process'.

The Management of Innovation also made much of competitiveness and dissension within management. The book was built around a basic distinction drawn between 'mechanistic' and 'organic' systems of management. This left open the question of what made the difference between the firms which saw the point of changing their management style and, sometimes, structure, and did so, however much trouble it took; those which saw the point and didn't; and those which saw no reason at all to change their ways. It wasn't a matter of size; two of the largest firms were also among the most innovative and flexible. Nor was there much to choose between those mostly concerned in large-batch production and those which had previously concentrated on bespoke ('jobbing') work or products highly dependent on advanced technology.

It was perhaps conceivable that an explanation might lie in what went on in the structure of management and supervision itself. People who work in firms are at one and the same time co-operators in a common enterprise and rivals for the material and intangible rewards of competition for advancement. The hierarchic order of rank and authority that prevails in them is at one and the same time an integral, unitary, control system and a career ladder. Matters are further complicated by the fact that we recognize both co-operative and competitive systems as necessary and reconcilable, to the point of obscuring what might otherwise be seen as the bounds of moral legitimacy. The 'rules of the game', in this sense, can differ quite sharply from one organization to another. It is also possible for the rules of the game to be challenged. It is this which makes for internal politics—a topic that surfaced, along with 'careerism', in every organization in the study. For careerism may also turn into, or reinforce, alliances with others who make the same kind of contribution to the organization and who, to that extent, share their interests and expectations about the distribution of resources and decision-making power. 'Internal politics' create means–end systems which are alternative and sometimes even discrepant with the publicly endorsed purposes and operating procedures of the organization.

Careerism and internal politics were visibly present in every one of the sixteen or so firms, English as well as Scottish, that the study eventually covered. Nor did they seem necessarily to resolve themselves either in orderly settlement or in reconciling differences by reconstructing the firm's policy.

I thought this idea worth following up, to the point of producing a paper, 'Micropolitics', which appeared in the same year as *The Management of Innovation* as a kind of supplement to it. It drew on the same materials as the book, with some additional points made by referring to the kind of academic politics more familiar to possible readers. While internal politics was a more or less constant element in the life of any large organization, I argued, it is—for the most part—when circumstances change fairly substantially that internal politics may work so as to ease the way for the internal changes necessary to accommodate the organization to its new situation. But a change in the circumstances of the firm may also threaten the standing and prospects of one or other contending group, and so prompt moves to defend its own position, resist any proposed changes, or positively obstruct them when they are made. Indeed, as I have already said, this second outcome was more often the case in the Scottish firms in the electronics study.

But afterthoughts were not enough, and it seemed worthwhile exploring these questions further in a quite different kind of organization—large, non-industrial, and preferably non-commercial—from the medium-sized industrial concerns in which I had developed these ideas. Almost by chance, an opportunity to do so arose in 1963, when I began what turned out to be a long-term study of the BBC. For reasons which are spelled out in the book's preface, *The BBC: Public Service and Private World*, which came out of it, was not published until 1977. All that needs to be said here is that it expanded on the ideas developed in the earlier study, and, I think, took them a good deal further, but did not amend them in any substantial way.

III

The convergence in organizational studies which appeared during a few years around 1960 was lost later on. On the other hand, while the picture of the industrial organization being developed during the 1960s had drawn further and further away from the conception embodied in scientific management or Weberian bureaucracy, it also became increasingly difficult to determine what positive conception was to take its place.

Things (the 'circumstantial postulate') became altogether different in the 1970s, and stayed so thereafter. The oil-producing countries—at long last—seized the opportunities offered by a *de facto* oligopoly, formed a cartel, quadrupled the price of oil, and precipitated the first of the confidence-sapping depressions of the last quarter-century. The effect of the resulting pressure on already fully loaded fiscal, banking, commercial, and manufacturing systems is too recent and too long-lasting to need underlining. Interest rates began a long-term rise, 'short-termism' became almost obligatory not only for business corporations but for banks, institutional investors, and governments. The high-road of successful entrepre-

neurship led to takeovers rather than to the cultivation of effective organizations, and this, in turn, led to the diversion of profits to dividends
which would boost share-prices and fend off corporate raiders, rather than
to investment of money and time in technical or organizational innovation. 'Reagonomics' ruled; inflation galloped; relative production costs
rose; and Japanese industry swept the board, with the newly industrialized
'Pacific Rim': Korea, Taiwan, Hong Kong, Singapore—and, now, South
China and Malaysia—following suit.

The first sign of the changed situation so far as organizational studies
were concerned was that there was some retreat from the wholesale rejection of the bureaucratic model. The publication of the full text of Weber's
Economy and Society in 1968 did something to rekindle interest in his
account of bureaucracy. A good deal of work was put into the search for
some overall principle which would either incorporate or supersede the
work of the previous two decades. The management structure of industrial and business organizations was pictured as determined by the prevailing technology, variously construed. Or the nature and structure of
organizations was of subordinate interest to, and dependent upon, the network of powerful interests or the overall governmental, business, and
industrial system of which it forms part and which controls its size, structure, and destiny—the argument of the once-influential 'corporate inter-
lock' school. Or, yet again, it was the creature of the 'institutional thought
structure' which prevails among the political, business, and administrative
leadership, whether at the local community, the national or—nowadays—
the international level. It seeks above all to block, cushion, or somehow
seek accommodation with adverse and competing circumstance or innovative and dissenting opinion so as to preserve the system and the context
of beliefs and values which sustains it—and by which it sees itself sustained. The list is by no means exhaustive.

Quite apart from the changes of scene within the areas of organizational
studies which are traditionally the province of sociology and social psychology, the reawakened interest of economists in what might be called
the 'internal economy' of the firm had also prompted fresh initiatives and
opened up new perspectives. There appeared a lengthy series of writings
by economists who wrestled with the seeming anomaly presented by
complex organizations set up in order to eliminate the costs of market
transactions between stages of production and even of distribution and
marketing—all this within an economic system presumed to reflect the
superior effectiveness of market transactions. A famous paper ('The

Nature of the Firm') by R. H. Coase which had appeared in 1937 is now regarded as having set this line of analysis going, although his ideas were not taken up until the 1960s, when the phenomenon of 'market failure' began to attract attention. Economic historians, too, extended their interest in the conceptual framework within which industrial capitalism had grown up to the changing institutional structure of the firm. A. D. Chandler's pioneering work, *Strategy and Structure*, which appeared in 1962, marked the beginning of what is now a well-established tradition of business history.

The trouble was that the rapid and simultaneous enrichment of organizational study from so many new and varied sources made life difficult for both established practitioners and new entrants. What had been a variegated but nevertheless interrelated and fairly coherent set of ideas about the nature of organization—which amounted to what might be called a recognized 'state of the art' body of knowledge—seemed to be drowned in a welter of disparate approaches and findings.

The combination of all these circumstances alone may well be partly responsible for the recrudescence of the hard-line managerialism which has manifested itself in recent years first in America and then in Britain and Europe—although there are, heaven knows, plenty of other explanations on offer for that development.

TOM BURNS

Edinburgh, 1994

PREFACE TO THE SECOND EDITION

The title, 'The Management of Innovation', has something of a didactic, textbook flavour about it. This was unintentional. The book attempts, not altogether successfully, to combine three kinds of critical study within the framework of an essay in organizational analysis.

I. THE ANALYSIS OF ORGANIZATIONS UNDER CONDITIONS OF RELATIVE STABILITY AND CHANGE

The core of all the twenty studies on which the book is based is the description and explanation of what happens when new and unfamiliar tasks are put upon industrial concerns organized for relatively stable conditions. When novelty and unfamiliarity in both market situation and technical information become the accepted order of things, a fundamentally different kind of management system becomes appropriate from that which applies to a relatively stable commercial and technical environment. This general thesis is set out in detail and argued at length in Chapters 4, 5, and 6. There is very little I would wish to add or to amend in this section, although, so far as I am able to judge from subsequent uses and references, there must be some wrongly placed emphases.

In the first place, the utility of the notions of 'mechanistic' and 'organic' management systems resides largely in their being related as dependent variables to the rate of 'environmental' change. 'Environmental', in this connection, refers to the technological basis of production and to the market situation; indeed, as I have argued elsewhere,* and as I think is implicit in Chapters 3 and 4, the increasing rate of technological change characteristic of the last generation could plausibly be regarded as a function of fundamental changes in the relation of production to consumption.

If the form of management is properly to be seen as dependent on the situation the concern is trying to meet, it follows that there is no single set of principles for 'good organization', an ideal type of management system which can serve as a model to which administrative practice should, or could in time, approximate. It follows also that there is an overriding

* T. Burns, 'The Sociology of Industry', in A. T. Welford (ed.) *Society*, London, Routledge & Kegan Paul, 1962 (p. 191).

management task in first interpreting correctly the market and technological situation, in terms of its instability or of the rate at which conditions are changing, and then designing the management system appropriate to the conditions, and making it work. 'Direction', as I have labelled this activity, is the distinctive task of managers-in-chief, and is discussed in the final chapters in terms of a critical examination of the role of the managing director; but properly speaking the description and explanation of the interpretative, directive functions of top management belong to the earlier section, since it is these functions that largely determine the effectiveness of the organization as a whole. The examination of the role of the chief executive, that is to say, is logically antecedent to the description and analysis of the organizational problems discussed in the middle section of the book (Chapters 7, 8, and 9). For it was failure at this level that produced the characteristic forms of management system encountered in most of the concerns—forms which, from the point of view of their overt constitution and purposes, were clearly inefficient and ineffective, although there are, as we shall see, other points of view from which they may be regarded as serving certain ends and performing certain functions in an entirely efficient manner.

For the individual, much of the importance of the difference between mechanistic and organic systems lies in the extent of his commitment to the working organization. Mechanistic systems (sc. 'bureaucracies') define his functions, together with the methods, responsibilities, and powers appropriate to them; in other words, however, this means that boundaries are set. That is to say, in being told what he has to attend to, and how, he is also told what he does not have to bother with, what is not his affair, what is not expected of him, what he can post elsewhere as the responsibility of others. In organic systems, the boundaries of feasible demands on the individual disappear. The greatest stress is placed on his regarding himself as fully implicated in the discharge of any task appearing over his horizon, as involved not merely in the exercise of a special competence but in commitment to the success of the concern's undertakings approximating somewhat to that of the doctor or scientist in the discharge of his professional functions.

In studying the electronics industry, I was occupied for the most part with concerns that had been started a generation or more ago, well within the time-period of the second phase of industrialization, and equipped at the outset with working organizations designed in accordance with mechanistic principles.

Enlarging the commitment of the individual to the concern in such a way as to admit of the adaptation of the working organization to its own larger commitment to the new situation confronting it, proved only partially possible to most firms. Indeed, the ideology of formal bureaucracy seemed so deeply ingrained in industrial management that the common reaction to unfamiliar and novel conditions was to redefine, in most precise and rigorous terms, the roles and working relationships obtaining within management along orthodox lines of 'organization charts' and 'organization manuals', and to reinforce the formal structure. In these concerns the effort to make the orthodox bureaucratic system work (because it was seen as the only possible mode of organization, and because the enlargement of commitments to the concern was abandoned as hopeless or never seriously contemplated) produced dysfunctional forms of the mechanistic system.

Three of these dysfunctional systems may be mentioned. All three were responses to the need for finding answers to new and unfamiliar problems and for making decisions in new circumstances of uncertainty. In the mechanistic system, the normal, prescribed, procedure for dealing with any matter lying outside the boundaries of anyone's functional responsibility is to refer it to the point in the system where such responsibility is known to reside, or failing that, to lay it before one's superior. If conditions are changing rapidly, in comparison with what obtained in the past, such episodes occur frequently; moreover, in many instances, the immediate superior has to put such matters higher up still. A sizeable volume of matters for solution and decision can thus find their way to the head of the concern. There can, and frequently does, develop a system by which a large number of executives find—or claim—that they can get matters settled only in direct consultation with the head of the concern.

So, in some concerns, there developed an ambiguous system of an official hierarchy of power and responsibility, and a clandestine or open system of pair relationships between the head of the concern and some dozens of persons at different positions below him in the management. The head of the concern would be overloaded with work, and senior managers, whose standing depended on the operation of the formal system, would feel aggrieved at being bypassed. The managing director would tell himself, or would bring in consultants to tell him, to delegate responsibility and decision-making. The organization chart would be redrawn. But, inevitably, this strategy promoted its own counter-measures from the beneficiaries of the latest system as the stream of novel and unfamiliar problems built up anew.

Disparities between the two systems had their effect on the operational effectiveness of the concern. The conflict between managers who saw their standing and prospects depending on the ascendancy of one system or the other deflected attention and effort into internal politics. Both processes bore heavily on the time and effective effort the head of the concern was free to apply to the tasks proper to his position, the more so because political moves focused on controlling access to him.

The preferred solution in other concerns was to grow more branches of the bureaucratic hierarchy. Most of the problems that appeared in the firms under discussion manifested themselves as difficulties in communication. These were met, typically, by creating special intermediaries and interpreters: methods engineers, standardization groups, contract managers, post-design engineers. Underlying this familiar strategy were two equally familiar characteristics of managerial thinking. The first is to look for the solution of a problem, especially a problem of communication, in 'bringing somebody in' to deal with it. A new job, or possibly a whole new department, may then be created, which depends for its survival on the perpetuation of the difficulty. The second attitude probably derived from the traditions of production management, which cannot bring itself to believe that a development engineer is doing the job he is paid for unless he is at a bench doing something with his hands, or a draughtsman doing his unless he is drawing at his drawing-board, and so on. Higher management in many firms are also worried when they find people moving about the works, when individuals they want are not 'in their place'. They cannot, again, bring themselves to trust subordinates to be occupied with matters directly related to their jobs when they are not demonstrably and physically 'on the job'. Their response, therefore, when there was an admitted need for 'better communication', was to tether functionaries to their posts and appoint persons who would specialize in 'liaison'.

The third kind of pathological response was encountered only rarely in the firms included in the study: the establishment of committees. It appeared sporadically in many of them, but was feared as the characteristic disease of government administration. It is nevertheless of particular interest since (as the word indeed makes clear) the committee is a traditional device whereby wider commitments, over and above those encapsulated, may be undertaken without enlarging the bounds of feasible demand made on individual functionaries and so upsetting the balance of commitments. The appearance of new types of task and unfamiliar problems may seem either to be a temporary phenomenon or to involve decisions, responsibilities, and powers of a kind that cannot properly be committed to one man or one department and handled by them. This is particularly the case in bureaucratic hierarchies in which most of the considerations, most of the time, are subordinated to the career structure afforded by the concern—a situation by no means confined to the civil service or even to universities. The aim is to overcome the difficulty caused by an apparent need for a job calling for unfamiliar responsibility or new powers by creating a super-person—a committee.

II. THE PLURALITY OF SOCIAL SYSTEMS WITHIN THE ORGANIZATION

The problem of analysis only begins, of course, with the prediction of two polar 'ideal types' of organization and the indication of a taxonomy of intermediate and 'dysfunctional' types. Two related, and in fact complementary, questions immediately pose themselves; and their form points the transition from the first level of study, concerned with organizational types and their relationship to the technical and economic environment, to the second, which bears on the interrelationship and mutual shaping of individual and organization.

The first question, which arises directly from the preceding discussion, is why some concerns—indeed, most of those which took part in the studies—did not change their management system from mechanistic to organic as the general context, technical and commercial, of their operations changed from relative stability to fairly rapid change. For they were all firms which had deliberately chosen these new commercial and technical fields. They were staffed with professionally competent and ambitious managers and industrial scientists, who saw themselves as seizing the opportunities and welcoming the stimulation that the new and rapidly expanding electronic industry presented.

The second question concerns the logical inconsequence of the term 'dysfunctional' when applied to social action. It is entirely appropriate to characterize management systems as dysfunctional when they are manifestly preventing, or hindering, or distracting industrial organizations from achieving the purposes for which they exist. But working communities are much more than management systems or organizations. As soon as the span of considerations we bring into play widens beyond that of organizational analysis, it is clear that 'dysfunctional' types of management system are developed and maintained in terms of other patterns of action. Considered in this different light, dysfunctional types of management system are seen to be entirely effective and appropriate parts of a social organization in which they serve rationally as means to specific ends, albeit not the ends of the organization.

Organizations are co-operative instrumental systems assembled out of the usable attributes of people. They are also places in which people compete for advancement. Similarly, members of a business concern are at one and the same time co-operators in a common enterprise and rivals for the material and tangible rewards of successful competition with each other. The hierarchical order of rank and power, realized in the organization chart, that prevails in all organizations is both a control system and a career ladder.

This dualism reflects the order prevailing in society at large. There is a sense in which a national society is organized as a co-operative system. Nevertheless, complex societies such as ours depend for their survival on maintaining a flow of the most highly qualified people to the top places in society where the best talents are most needed. Accordingly, a complex system of educational and occupational promotions open to merit has been set up. But beyond this it is essential for every member of the society to enter the race and compete as best he can, and to regard success in these terms as one of the highest personal goals of his life. This indoctrination is usually fairly successfully carried out, especially because the distribution of rights and privileges accords more and more with the social position achieved in this way.

However, in the nature of things, few people out of the total numbers who have to try can actually succeed. Thus a Western society is composed of people almost all of whom are frequently confronted with the possibility of failure. In many cases, therefore, individuals will try to increase their own chances of success by illegitimate means and will impute the same desire to others who are, or may one day be, in competition with them. All social milieux in which such competition occurs have a code of rules, explicit and implicit, that distinguish illegitimate behaviour from legitimate, and apply sanction of some kind to all transgressors. Moral codes of this kind are specific to individual occupational milieux, and even to individual organizations; overt references to the failures and deficiencies of one's rivals, of the kind which are unexceptionable in the clothing trade, would damage oneself, rather than rivals, in the B.B.C., although, if we are to believe C. P. Snow, they may also go unquestioned in the less genteel social climate of a Cambridge college. The existence of such codes, and of such definitions, bespeaks the presence of a specific system of behaviour, with its own normative rules governing relationships—i.e. of a social system, existing in juxtaposition with the social system represented by the formal organization, with its own norms and relationships.

But there are also, frequently, disputes about the criteria that determine success, about the justice or logic with which they are applied, and about the relevance of the criteria of career success to the capacity of the organization or society as a whole to survive. In addition, the ends and means formulated by previous or existing leaders as goals and procedures for the organization as a co-operative system may be challenged. In short, the rules of the game being played in the two social systems I have already mentioned—the formal organization and the career system—may be disputed. Major decisions about changing the rules of the game usually require much antecedent search and discussion, not so much to accumulate information as to test out and align consensus; decisions about 'policy matters', therefore, ordinarily involve the formation of groups and sections concerned with the advocacy of one or other course of action. But such groupings are more enduring than this suggests. The individual may be, and usually is, concerned to extend the control he has over his own situation and prospects to increase the value of the resource which he represents to the organization. He may be, and often is, able to increase his personal power by attaching himself to parties or sections of people who represent the same kind of resource, and wish to enhance its exchange value, or to cabals who seek to control or influence the exercise of patronage in the organization. Such formations have to do with what are commonly, and rightly, regarded as the 'internal politics' of organizations. The issues in which the interest groups thus formed are involved arise from conflicting demands on the rewards of the enterprise, its uncommitted resources (the allocation of new capital), the direction of the activities of others (the authority to propose or sanction action by other departments, groups, or individuals), and patronage (promotions, new appointments, the distribution of rights and privileges).

The systems of relationships and the kinds of behaviour which can be designed 'political' are also, of course, subject to the same kinds of normative regulation that obtain for the formal organization and the career system. It is this which enables us to recognize political activity when we see it. Normally, each side in any conflict called by observers 'political' claims to speak in the interests of the organization as a whole. In managerial as well as other legislatures, both sides identify their interests with those of the community as a whole. It is only backstage, so to speak, that the imputations of empire-building, log-rolling, and obstructionism occur.

The terminological division is particularly marked in universities, where mutually exclusive frames of reference exist for discussion in faculty meetings or committees, and in clubs, common-rooms, and parties. In the first, the only legitimate reference is to the needs of students and to the advancement of approved branches of learning. Allegations, or even hints, of careerism or of pressure-group politics are entirely improper in faculty meetings, yet may be entirely acceptable and legitimate in other settings. Indeed, in certain gatherings it might be imprudent to the point of social suicide to impute higher motives than self-interest and aggrandisement to reformers and claimants for larger funds or more staff.

These three kinds of commitment that I have specified, and the three kinds of social system that are built out of action committed in these ways, are not exhaustive. Nor is the tripartite set of commitments self-balancing for the individual or the tripartite set of systems in equilibrium. Beside commitments to the concern, to 'political' groups, and to his own career prospects, each member of a concern is involved in a multiplicity of relationships. Some arise out of incompatibilities and compatibilities of social origin, sub-culture, and age. Others are generated by the encounters that are governed, or seem to be governed, by a desire for the comfort of friendship or the satisfaction of popularity and personal esteem, or those other rewards which come from inspiring respect, apprehension, or alarm. All relationships of this sociable kind, since they represent social values, involve the parties in commitments. Each such commitment may be described as an investment of part of the social self in a value, an investment which carries with it the possibility of reward and the risk of loss. There is, so to speak, a quantitative variation in the amount invested—in the degree of commitment. Moreover, the total investment of one person in working, social, political, and career commitments may vary with such extrinsic commitments in a complementary fashion, as the Jungian polarity of extraversion-introversion originally presumed. There is no evidence to suggest that every individual human ego has the same amount of 'philopsychic' capital to invest.

However, those commitments that I have designated as political and careerist bear directly, whatever their strength, on the commitment of the individual to the working organization, and to its command of him as a resource. They do so largely, not merely because of their significance as commitments, but also because they are, as I have already suggested, alternatives modes of self-advancement or self-preservation to that which is contained in membership of the co-operative system of the working organization. Neither political nor career preoccupations operate overtly, or even, in some cases, consciously. They give rise to intricate manoeuvres and counter-moves, all of them expressed through decisions, or in discussions about decisions, concerning the organization and the policies of the firm. Since sectional interests and preoccupations with advancement only display themselves in terms of the working organization, that organization becomes more or less adjusted to serving the ends of the political and career system rather than those of the concern.

III. ORGANIZATIONAL DYNAMICS

The information all these studies provided, in interviews and meetings, in casual remarks or in lengthy expositions and revelations conveyed during mealtime or evening conversations, in the observation of everyday routines and happenings in offices, laboratories, and workshops, is information about actions and happenings. When the chronological sequence of notes and recordings becomes analysed as research data, the actions and happenings can be organized in terms both of the kind of commitments entered into by people and of the systematic relationships between aspects of actions and events which appear as one looks for their organizational significance, their political significance, or their career significance. This ambiguity in the analytical connotations of our information is not, of course, merely the product of applying some unconnected theoretical apparatus to facts that reside in an ontologically distinct category of existence. The ambiguity attaches to the actions and events themselves; they are equally the constituent elements of what we perceive as personalities, on the one hand, and as organizations, political systems, or status hierarchies (and as co-operation, conflict, or careerism), on the other. Moreover—and here one returns to almost platitudinous common-sense—each set of derivatives, individual identities, and institutional systems, is the creation of the other. The familiar facts of ecology pursue us, in short, into the immaterial circumstances of individual existence within an organized social environment, an observation which is at least as old as Hegel, but which is too slippery, apparently, for our habits of thought, founded as they are on the need to cope with a world of material techniques, of artifacts, and of 'objective', narrative history which the nineteenth century bequeathed us.

We are, I believe, closer to the study of the social world as a process,

---

* For a preliminary exercise in 'organizational dynamics', see T. Burns, 'Ambiguity and Identity', as yet unpublished.

instead of an anatomy frozen into 'structured' immobility; closer to the identification of the abstraction, society, with the empirical fact, behaviour, if we accept the essential ambiguity of social experience and organize interpretation of it in dynamic rather than in structural terms.* It is in this way, by perceiving behaviour as a medium of the constant interplay and mutual redefinition of individual identities and social institutions, that it is possible to begin to grasp the nature of the changes, developments, and historical processes through which we move and which we help to create.

There is no section or set statement in this book concerned with 'organizational dynamics', which is a more recent preoccupation of mine. Yet the research experience on which it drew provided the insights that demanded this kind of analytical treatment.

The first part of the book is directly concerned with identifying industrial organization as the product of the actually developing change processes in society and of the purposes and commitments of entrepreneurs and managers. What I have called the technological system itself is treated in the same way. And not only the 'dysfunctional' but also the two ideal types of management system are presented as the outcome of the responses of managers to their commercial and technical situation and the definitions of tasks and relationships which they accepted or rejected. At the level of organizational analysis, it is the dynamics of situations and systems that are, I think, kept in mind throughout. What is not pursued in this book is the way in which the systems of internal politics and careers have developed their special character and their special importance, and, equally, how preoccupation with internal politics and with careerism has increased as a consequence of the general change in the technical, commercial, and organizational context of managements' existence.

Like almost every other economic, political, and social organization that exists to devise and execute action within the framework of aspirations and institutional norms prescribed by others, or by society at large, an industrial undertaking forms a hierarchy of rank and power, more or less clearly defined and ascribed to each individual member by superiors. However, this hierarchy of subordination has to be underwritten by the beliefs of all members of the organization as legitimate and right. Superiors may not be right in everything they do, but are regarded as acting 'within their rights' to do it. In the past, the right of superiors to their rank and powers was hardly questioned outside the programmes of syndicalist and anarchist political groups, except in rare cases of demonstrable and disastrous incapacity. More recently, as production and the market have moved into a fundamentally unstable relationship, and as the stream of technical innovation has quickened, the legitimacy of the hierarchical pyramid of management bureaucracy has been threatened by the sheer volume of novel tasks and problems confronting industrial concerns. Manufacturers of durable consumer goods, such as cars, or industries founded on new techniques, such as electronics, have been most exposed to this threat. New and unfamiliar tasks and problems create situations and demands for information and action to meet them that are often incompatible with the presumption of the traditional hierarchy. To be an older man in any industrial concern used to mean the possession of local knowledge of the organization, of the plant and its resources, of the specialized equipment, of the capacities, limitations in skill, initiative, and commitment of labour, which made him more valuable, more knowledgeable, more fitted for seniority than younger men. Given reasonable equivalence in intelligence and qualifications, to be older meant that one was more effective and better qualified. But in the new situation of technical and commercial change, this basis of authority has become invalidated.

It is not merely that chief executives, and even heads of industrial laboratories, would confess that they found it impossible or difficult to grasp the vocabulary or implications of the technical information and the skills for which their juniors had been recruited into the organization. Marketing procedures and office management are undergoing equally rapid changes, and are increasingly invaded by operational research, by statistical procedures, by computing methods, and by information requiring a technical understanding of economics, all of which are utterly outside the training and experience of older, including senior, management.

The consequences of this alteration in the circumstances and value attached to seniority are visible in what seems to be an intensification of concern with internal politics and careerism. If superior technical skill and information, and the initiative in ideas and projects, have to be abdicated to the under-forties, then political power or, more specifically, the judicial monitoring, appellate role of senior executives may be reinforced by way of compensation. There is a sense in which the techniques of reporting, of working parties and internal investigations, and, particularly, of organizing the discussion of new ideas in the debating forum of 'presentation' sessions, has meant the sharpening, clarification, and admission of political divisions and conflicts within organizations; a development that has imported the stimulation of 'healthy rivalry' and 'constructive criticism' but that also clearly strengthens the authority of senior executives in whom the eventual power resides to choose, decide, approve, or veto. Individually and collectively, senior status in the hierarchy is validated increasingly by the role of judge and court of appeal and by senatorial remoteness from the contentions and partisan views of middle and junior management.

The signs of the consequences these changes have had on the career system are rather more publicly manifest. The contributory pressures to change are also more familiar. Post-war reform of the educational system has spread the assumption that the more intelligent and capable children are now effectively separated out and funnelled into grammar schools and that, of these, the cream separates itself out and passes through the universities and technical colleges. New managerial talent, therefore, is thought of as confined to graduates. So the field for corporations to pick from is ready graded. It is also smaller. Furthermore, there is a new industrial situation. The economy as a whole has worked at greater pressure than before the war. There is also the insecure balance between production and market and the concomitant acceleration of technical innovation we have already discussed. All three factors tend to raise the demand for managers, specialists, and technologists both in absolute numbers and in proportion to operatives.

Competition among employers for management recruits has raised starting salaries; but since investment in a recruit has to mature for many years before yielding a return, it is necessary both to select the best and to hold on to them. While no commercial enterprise selects its recruits by a system of open examination as formidable as that of Civil Service Entrance, much use is made of selection techniques developed during World War II in the choice of candidates for officer training courses. And careers thus begun are presented as enclosed for life within the corporation. Indeed, in their endeavours to attract talented candidates, corporations advertise careers rather than posts.

Inevitably, therefore, the career system has assumed greater importance. Business corporations are confronted with the need to give increased prominence to appointments and promotion. The more prominence given to career success, the more has preoccupation with it increased. The style and manner believed to match promise for elevation are more eagerly cultivated; favourable task positions and tasks that may catch the boss's eye are more assiduously sought, and the best routes to the top more openly canvassed. And such manoeuvres are more frequently read into the conduct of rivals.

Most corporations have taken steps to control the career system; at the very least, they have had to try to improve the chances of success being seen to depend on performance and capacities that are generally acknowledged as deserving. Management training courses elucidate the range of managerial tasks and specify managerial techniques and personal qualities. Conferences promote commitment to the world of management and give social, economic, political, intellectual, and sometimes religious point to its workaday activities; representative figures from among the successful are also presented to the rank-and-file of management to speak to them in broadly philosophical but brief terms of the ramifications of the corporation's interests, its concern for their welfare, its view of the future, and its hopes for them. Some corporations subscribe to the idea of job-rotation, by which managers of promise are posted to functionally remote parts of the enterprise at short notice in order to prove their quality, widen their experience, and distinguish abilities based on knowledge local to a department and technique from those based on skill and coolness in manipulating resources. And there is 'management development', which involves plotting the career paths of selected junior managers for some years ahead, placing them in positions in which they can encounter profitable opportunities for successful managerial exploits, conducting them through the range of functional dependencies of board policy, and dispatching them to the more expensive residential courses for managers provided nowadays by industrial federations or the older universities.

At this point organizational requirements are being visibly subordinated to those of the career system. Admittedly, this is in the hope of substantial returns in the form of better selected, more experienced, and more informed higher management; but to the individual manager involved as the object, or the planner, or the facilitator, or the displaced victim of all these contrivances, it seems that the competitive career system has become not only the paramount element in the concern but a game of snakes and ladders played by one or two powerful individuals with the sanction and approving interest of the board, and that it has ousted the realities of the co-operative system of management, however much 'leadership'—the ability to promote co-operation and loyalty to the firm's interests—is insisted on as the fundamental managerial talent.

The course that these developments have followed has not been laid down by some occult and sinister design. There has been a natural, though not perhaps inevitable, deflection from adjustment to new situations into courses of policy that will preserve the image senior management have of themselves that will reinforce their authority at a time of innovation and of new ideas and new men, and that will present an appearance of bold and zealous reform to meet new circumstances while nevertheless holding on to the essentials of the traditional system which they know and which brought them to the top. Constitutional and organizational changes in many organizations are liable to demonstrate these points of influence, and many changes are subject to them.

The attention paid to the career system arises from the changes in the managerial labour market. But the central position occupied in corporation thinking about organization by selection, appointment, promotion, and career planning has fostered the notion that, as one senior manager declared, 'organization really means getting hold of the best men and training them'. The implications of the prescription are, first, that the positions at the top are filled, or were created, by men with qualities that can be matched in their successors only if the utmost diligence is exercised in their choice and grooming over long periods of years, and, second, that the main managerial responsibilities of designing, and improving the design of, the management system, and of seeing that individuals are provided with the conditions under which they can reach decisions and act most effectively, are discharged by 'getting a good man in', and only not discharged because there are no 'good men' available.

IV. THE NEW INDUSTRIALISM

The whole book is, at different levels, a critical study of an institutional area that is regarded as of some importance to the welfare of society. I hope that it conveys also some indication of other significances than this. Our society, and we as members of it, are the creatures of industrialism. Historically, the industrial system has been seen as imposing its own structure of relationships on the people who work for it, on their dependents, and, eventually, on all members of society who are, so to speak, 'processed' by its needs for human resources and, now, into conformity with its needs for consumption of its products. This is, though true, only half the truth; for industry has not so far escaped, as science has, from the constraints imposed by its evaluation in instrumental terms. But even though the special imprint that industrialism puts on society is the product of patterns and combinations of action which are assuredly more than the outcome of the social and technical order of industry itself, the dynamics and material consequences of industrialism have altered in extent, in character, and in the degree to which they have involved the lives of working or non-working members of society. And the kind of processing effect has altered with its own institutional character, along with the technological system which both supports and is supported by industry.

The reunion of industry and science through the new technology effects a radically different, much more intimate combination of forces than anything that has obtained before. The industrialism we now have, or are getting, is a new industrialism; its effect on people inside and outside the immediate world of industry is substantially different from what we, in the oldest industrial country in the world, have come to accept as customary, maximal, and acceptable. The electronics industry is itself one of the first creatures of the new combination of science and industry, one of the earlier manifestations of the new industrialism. It exhibits in an elementary and articulate fashion the novel characteristics of organizational and work setting, the kinds of response that the people first and most directly involved make to its demands, and—more particularly—the way in which they regard it. This book is necessarily a report of their findings. In many respects, the greater part of it is no more than a systematic rendering of the observations made about their own experience by industrial scientists, company directors, draughtsmen, technicians, and skilled workers, production engineers, and foremen. It is they who, in different, more direct, and less elaborate terms, distinguished the two 'ideal types' of management system; it is they who first outlined the internal politics of their firms and suggested the ways in which management structures and policy decisions were often more easily comprehended if they were seen as consequences of political moves made by this or that section; it was in discussion with them that the significance of the commitment of individuals to their own careers, and the fact and the meaning of the increasing preoccupation with career advancement first became plain. Some justice has been done to their contribution in the lengthy quotations from interviews that occur in one or two chapters, but not enough. Now, some years after writing the book, and with the added experience of a good deal of subsequent research, I am more than ever impressed with the extraordinary gap that exists between the perceptiveness, intellectual grasp, and technical competence of the people who work in industrial concerns, and the cumbrous, primitive, and belittling nature of the administrative structures by which they direct their efforts, and of the constraints they see fit to impose on their thinking and liberty of action.

TOM BURNS
Edinburgh
April 1966

CHAPTER 1

Introduction

All the research reported in this book arose out of an attempt, some years ago, to study an industrial concern as a 'community of people at work', that is, in much the same terms one would use in a study of conduct and relationships in a village, an urban neighbourhood, or a small primitive community. This aim was never realized, because it soon became evident that the social structure of the factory interlocked with, and often mirrored, that of the small isolated town in which it was situated. The wider study which then appeared necessary was not practicable and the enquiry petered out rather inconclusively, assuming its present significance only in the context of later studies.

At this time, the concern (the rayon mill described in Chapter 5) was growing and commercially prosperous. But two sets of circumstances which the study revealed did not seem easy to square with first-hand knowledge of other firms and with the conceptions of management available in the literature. Partly because of the lead given from the head office in London, the functions of each manager and worker were clearly specified; they were expected to follow, and did follow, the instructions which issued in a steady flow from the general manager and down through the hierarchy. Yet the system, lubricated by a certain paternalism, worked smoothly and economically, and there was no evidence that any individual felt aggrieved or belittled.

The other feature of interest lay in the comparative impotence of the Research and Development Laboratory. It was formally responsible for solving problems and curing faults in the process other than those which could be tended by people on the spot, for improving the existing process and products, and for introducing new products or methods. But its activities were regarded with much suspicion and some hostility by many production managers and supervisors; its studies were repetitive and often inconclusive; it was very largely occupied with finding answers to enquiries from the London office which arrived almost daily, and large arrears of which, at that time, had accumulated.

Very soon an opportunity presented itself of carrying out a similar study of the organization of an engineering concern with very large development interests. The wholly different conditions in which management acted, and the different codes of conduct and beliefs which individual managers brought to their jobs, were abundantly clear at the very beginning. As in the first concern, the study began with a series of interviews with managers and foremen, the principal purpose of which was to obtain descriptions of the jobs performed by individuals and the way in which they fitted in with others. After the first few such interviews a pattern appeared in them which was entirely unanticipated. The usual procedure was that after listening to the researchers, explaining his presence in the factory and his present purpose the informant would say, 'Well, to make all this clear, I'd better start from the beginning.' He would then proceed to give an account of his career in the firm, and of the activities and duties characteristic of the positions he had filled. This account was commonly lucid, well-organized, and informative, but would stop short at a point some months earlier. The question about his present functions, and whom they affected would then be framed again, rather more pointedly. There would be a pause. He would then explain, equally lucidly, what he would be doing when the present emergency had passed or the current reorganization or new development had matured, and his part of the concern could settle down to work as it was now planned.

Later, it became evident that ranks in the hierarchy of management as well as functions were ill-defined, and that this was so because of the deliberate policy of the head of the concern. At this time, the most obvious consequence of this state of affairs was a pervasive sense of insecurity which was openly discussed by some managers and was also evident in individual conduct and in the formation of cliques and cabals.[1] Yet there was also the striking fact of the concern's commercial and technical success. Was there a causal connexion between the insecurity and stress displayed by individuals and the concern's effectiveness? An American study,[2] published about this time, suggested that there might be. Yet many of the actions arising from anxiety about career prospects and status were so clearly dissociated from the concern's tasks, even running counter to their accomplishment, and so much energy was consumed in internal politics, that it still seemed more plausible to regard insecurity, and the conduct to which it gave rise, as defects of the management system rather than its mainspring. Possibly, though, these defects were an inevitable concomitant of industrial change in the present state of our knowledge of organization.

This, at all events, was the view of the head of the concern. An organization chart was inapplicable, he believed, to the structure of management in the concern—it was 'probably a dangerous way of thinking about the way any industrial concern worked'. The first requirement of the management system was that it should make the fullest use of the capacities of its members; any man's job, therefore, should be as little defined as possible, so as to allow it to expand or contract in accordance with his special abilities. Any anxieties and frictions that might be generated were an inevitable circumstance of life as it is, and one could not 'manage them out of the organization'— not, at least, without neglecting or damaging some more vital interest.

Further study suggested, however, that 'initiative' no less than 'insecurity' and 'stress' might be dependent on the way in which management organized itself to carry out its task. The adaptation of *relationships* between individuals, rather than of individuals themselves, towards the requirements of the technical and commercial tasks of the firm became the focal point of the broader study which was then initiated, with the financial backing of the Department of Scientific and Industrial Research and in partnership with G. M. Stalker.

THE SCOTTISH STUDY

This next phase of research was an attempt to analyse the experience of the firms which entered the Scottish Council's 'electronics scheme'. The origins and working of this scheme, which itself threw a good deal of light on the more general considerations of 'the task of the firm', is described in Chapter 3 (pp. 45–51). It was intended to facilitate the entry of Scottish firms with the right resources into electronics development work and, eventually, manufacture.

The Scottish Council (Development and Industry) is a voluntary body supported financially by industrial firms, local government bodies, and trade unions, and works in close touch with the Scottish Home Department and the Board of Trade. It has actively encouraged the growth in Scotland of industries using newer techniques. The declared purpose of the electronics scheme is to enable firms to acquire new technical resources and exploit them in commercial fields reasonably familiar to them. It is to this end that the firms are helped to build up laboratory teams on the basis of suitable contracts provided by defence ministries.

For our part, we hoped to be able to observe how management systems changed in accordance with changes in the technical and commercial tasks of the firm, especially the substantial changes in the rate of technical advance which new interests in electronics development and application would mean.

Study of the firms in the scheme was later supplemented by ancillary studies of the experience of firms in other industrial fields, but moving in much the same way from a relatively stable situation to one dominated by technical progress.

The major consideration for most firms entering the scheme was fear of shrinking markets or of keener competition in a static market; only one or two seemed prompted by an expansionist urge and the attraction of enterprise in new fields. A second distinction revealed itself between firms which negotiated a development contract before engaging a laboratory team, and those which began by investing in people who might be expected to produce ideas for development. Following roughly the same lines of division, a third distinction was visible between firms which confined the activities of their laboratory teams to work on defence contracts or on improving products developed elsewhere (to the extent of refusing to invest their own capital in development), and those prepared to exploit the team as a technical resource.

No firm attempted to match its technical growth with a comparable expansion of sales activities; in particular, no attempts were made at organized and thorough exploration of user needs for products which firms thought it possible to develop, or even for those which they had developed.

Most of the Scottish firms failed to realize their expectations. In half the cases, laboratory groups were disbanded or disrupted by the resignation of their leaders. Others were converted into test departments, 'trouble-shooting' teams, or production departments. Common to all predicaments was, first, the determined effort from the outset to keep the laboratory group as separate as possible from the rest of the organization; second, the appearance of conflicts for power, and over the privileged status of laboratory engineers; and third, the conversion of management problems into terms of personalities—to treat difficulties as really caused by the ignorance, stupidity or obstructiveness of the other side. These failures were interpreted by us as an inability to adapt the management system to the form appropriate to conditions of more rapid technical and commercial change.

There seemed to be two divergent systems of management practice. Neither was fully and consistently applied in any firm, although there was a clear division between those managements which adhered generally to the one, and those which followed the other. Neither system was openly and consciously employed as an instrument of policy, although many beliefs and empirical methods associated with one or the other were expressed. One system, to which we gave the name 'mechanistic', appeared to be appropriate to an enterprise operating under relatively stable conditions. The other, 'organic', appeared to be required for conditions of change. In terms of 'ideal types' their principal characteristics are briefly these:

In mechanistic systems the problems and tasks facing the concern as a whole are broken down into specialisms. Each individual pursues his task as something distinct from the real tasks of the concern as a whole, as if it were the subject of a sub-contract. 'Somebody at the top' is responsible for seeing to its relevance. The technical methods, duties, and powers attached to each functional role are precisely defined. Interaction within management tends to be vertical, i.e., between superior and subordinate. Operations and working behaviour are governed by instructions and decisions issued by superiors. This command hierarchy is maintained by the implicit assumption that all knowledge about the situation of the firm and its tasks is, or should be, available only to the head of the firm. Management, often visualized as the complex hierarchy familiar in organization charts, operates a simple control system, with information flowing up through a succession of filters, and decisions and instructions flowing downwards through a succession of amplifiers.

Organic systems are adapted to unstable conditions, when problems and requirements for action arise which cannot be broken down and distributed among specialist roles within a clearly defined hierarchy. Individuals have to perform their special tasks in the light of their knowledge of the tasks of the firm as a whole. Jobs lose much of their formal definition in terms of methods, duties, and powers, which have to be redefined continually by interaction with others participating in a task. Interaction runs laterally as much as vertically. Communication between people of different ranks tends to resemble lateral consultation rather than vertical command. Omniscience can no longer be imputed to the head of the concern.

The central problem of the Scottish study appeared to be why the working organization of a concern did not change its system from 'mechanistic' to 'organic' as its circumstances changed with entry into new commercial and technical fields. The answer which suggested itself was that every single person in a firm not only is (a) a member of a working organization, but also (b) a member of a group with sectional interests in conflict with those of other groups, and (c) one individual among many to whom the rank they occupy and the prestige attaching to them are matters of deep concern. Looked at in another way, any firm contains not only a working organization but a political system and a status structure. In the case of the firms we studied, the existing political system and status structure were threatened by the advent of a new laboratory group. Especially, the technical information available to the newcomers, which was a valuable business resource, was used or regarded as an instrument for political control; and laboratory engineers claimed, or were regarded as claiming, élite status within the organization.

Neither political or status preoccupations operated overtly, or even consciously; they gave rise to intricate manoeuvres and counter-moves, all of them expressed through decisions, or discussions about decisions, concerning the internal structure and the policies of the firm. Since political and status conflicts only came into the open in terms of the working organization, that organization became adjusted to serving the ends of the political and status system of the concern rather than its own.

The individual manager became absorbed in conflicts over power and status because they presented him with interests and problems more immediately important to him and more easily comprehended than those raised by the new organizational milieu and its unlimited liabilities. For increases in the rate of technical and commercial change meant more problems, more unfamiliar information, a wider range of work relationships, and heavier mental and emotional commitments. Many found it impossible to accept such conditions for their occupational lives. To keep their commitments limited meant either gaining more control over their personal situation or claiming exemption because of special conditions attached to their status. These purposes involved manoeuvres which persistently ran counter to the development of an organic system, and raised issues which could only be resolved by a reversion to a mechanistic system.

The Scottish study developed eventually into two complementary accounts of the ways in which the adaptation of management systems to conditions of change was impeded or thwarted. In one set of terms, the failure to adapt was attributed to the strength of former political and status structures. In other terms, the failure was seen as the consequence of an implicit resistance among individual members of concerns to the growth of commitments in their occupational existence at the expense of the rest of their lives.

THE ENGLISH STUDY

During the winter of 1955–6, the authors read papers dealing with some of the general findings of the Scottish study at a number of meetings. One of these was attended by senior officials of the Ministry of Supply. In later conversations with Burns, they suggested that major firms in the electronics industry in England might like to have an opportunity of hearing about the Scottish study and of discussing its implications for their own concerns. This suggestion led to a meeting in November 1957 at the Ministry of Supply, which was attended by managing directors and other senior members of eleven English firms, and by government officials.

While this discussion made it clear that the problems discussed in the summary report of the Scottish study were not unfamiliar, the ways in which the problems revealed themselves in different firms, and the responses and actions which they had evoked, were varied and idiosyncratic. Burns was therefore invited to make a brief study of each firm. Each of these studies concentrated on two topics: the management difficulties which seemed peculiar to firms engaged in rapid technical progress, and the particular problem of getting laboratory groups on the one hand (research—development—design) to work effectively with production and sales groups on the other.

The survey of English firms was completed in the first half of 1958 and the findings reported to the eight firms which had participated, out of eleven invited to do so. A general report was also distributed, and discussed at a one-day conference of the heads of firms and of government officials held at the Department of Scientific and Industrial Research in July 1958.

The eight English firms which eventually took part in the survey were not only much larger but much more committed to electronics development and manufacture than were the firms of the Scottish study, which were in the earliest stages of their careers in electronics. The situations available for study were more complicated; they were also more intimately related to the commercial and industrial destinies of the firms and to the lives of the people in them.

There was, for example, much more variety in the kind of group within the firm affected by an acceleration in the rate of technical change, and in the responses to change made by different firms. In firms which operated consciously on organic lines, changes from any direction were regarded as what they manifestly were—circumstances which affected every part of the firm and everybody's job, in some way. Organizational changes, additional tasks, and growth in any particular direction tended to be seen as the concerted response of the firm to a new situation; although debate and conflict were present, they were manifestly present and could be treated as part of the new situation to be reckoned with. In firms which operated according to mechanistic principles, the response to change was usually to create a new group, or to reconstitute the existing structure, or to expand an existing group which would be largely responsible for meeting the new situation, and so 'not disrupt the existing organization'.

This latter response, which in the Scottish firms characteristically led to the segregation of the new development team from the rest of management, was now visible in the way some firms dealt with big changes in market conditions. A Head Office sales department, or a new sales forecasting and market study group, might be created. Management might be reconstructed on product division lines, so as to extend the control of sales over the activities of the firm. Engineers might be recruited from development laboratories, or directorships offered to men of outstanding reputation from other firms. More significantly still, a new technical departure might be made the province of a newly created laboratory group independent of the laboratory concerned with the obsolescent techniques. In such cases, the confinement to a prescribed section of its organization of the total response of the firm to change meant that for the rest of the firm the challenge of the new situation became instead a threat offered by the 'new men' to the power, standing and career prospects they had hitherto enjoyed. This was especially the case with development engineers. Previously the element in every firm which had been identified with expansion and innovating change, they now saw their leading role passing—in part—to sales. The development-production conflicts typical of the Scottish study were overshadowed in the English firms by sales-development conflicts, by the resistance of the professional innovators to an innovating change.

Political conflict appeared to be clearly related to the particularism which was fostered by the separating out of the tasks of the firm according to specialist functions. Given a mechanistic system, changes of all kinds, including expansion, continually threw up new institutions within the firm which were intended to carry the whole of a new defined task and which themselves engendered political problems.

The conceptions of mechanistic and organic management have also proved useful in analysing the arrangements made inside firms for passing work through from the earliest stages of development to final manufacture. The tendency to regard the whole process as an articulated series of separate specialist functions made for the creation of 'hand over' frontiers between departments and for language barriers; it also went with a predilection for tethering functionaries to their posts. The need for communications beyond the formal transmission of instructions and drawings led to the appointment of liaison specialists—interpreters whose job was to move across the linguistic and functional frontiers and to act as intermediaries between the people 'getting on with the job'. Organic systems recognized the supreme importance of common languages and of each functionary's being able to seek out and interpret for himself the information he needed. The fewer distinguishable stages, the fewer interpreters and intermediaries, the more effectively were designs passed through the system.

Many of the insights generated by the English study were suggested in the first place by the distinctive response made by different concerns to a major change in market conditions as against techniques, as was the case with the firms in the Scottish study. The decline in government work and the increased emphasis on selling in the so-called 'commercial market' affected all concerns in the same way, although to a different extent. The first observable distinction was between the firms which saw that a sales function had been discharged by the laboratory engineer working on government development contracts, and that a similar role was equally necessary with commercial users, and those which overlooked this sales function in connexion with defence ministries and regarded market exploration and development as the province of salesmen. There were a number of aspects of this difference. Some concerns had always been wary of committing themselves too heavily to government work; others had allowed themselves to become educated into commercial unfitness by too complete a dependence on defence contracts. In general, it could be said that the first kind of firm tended to regard the market as a source of design ideas which the firm then attempted to realize, the second kind as a sink into which should be poured applications of techniques developed within the firm. Successful manufacturers of domestic radio and television receivers offered the most striking demonstration of the first principle. So much so, that in these firms not only the management system but the way in which individuals' jobs were defined, and the code of conduct prevailing in the concern, seemed to be generated by constant preoccupation with the market on the part of every member of management.

The differences between the two kinds of management system seemed to resolve themselves into differences in the kind of relationships which prevail between members of the organization, whether of the same or of different rank, and thus into the kinds of behaviour which members of an organization treat as appropriate in their dealings with each other. It was possible to distinguish various modes of behaviour used by individuals according to a single dimension of conduct: the bounds set to what—in the way either of requests, instructions, or of considerations and information—the individual would regard as feasible, acceptable, worth taking into account, and so forth. The observable way in which people in a concern dealt with each other—the code of conduct —could therefore be regarded as the most important element in a concern's organization, given the structure of the management hierarchy and the skills and other resources at its disposal. It expresses the framework of beliefs which decision-making invokes. In a realistic, operational sense, it *is* the organization.

In so far as differences in the obligations and rights attaching to different status within the concern are disputed, and in so far as the allocation of control over resources becomes a matter of political conflict, the style of conduct employed by the contending parties shows differences. That is to say, each side has differing beliefs about what considerations should enter into decisions, and about what are the feasible limits of the demands for action which may be made of themselves and which they may make on others. Conflicts thus wear the aspect of ideological disputes, whether these are conducted in overt terms or are implicit. The head of the concern enters at this point as a key figure who, in manifest or latent ways, denotes the code of conduct which should obtain.

TECHNICAL PROGRESS AND THE OCCUPATIONAL SELF

Organic systems are those which are best adapted to conditions of change. By common consent, such conditions are at present affecting a widening sector of industrial and occupational life. The code of conduct characteristic of organic systems—those better fitted to survive and grow in changing conditions—comprehends more eventualities than that necessary in concerns under stable conditions. More information and considerations enter into decisions, the limits of feasible action are set more widely.

The extension of the boundaries of feasible action and pertinent consideration makes for a fuller implication of the individual in his occupational role. As the pace of change, especially technical change, accelerates, and as the organic systems better equipped to survive under these conditions also expand, the occupational activities of the individual assume greater and greater importance within his life. This is in keeping with the commonly observed tendency for occupational status to assume an increasingly dominant influence over the location of individuals in British society. But it also denotes a greater subjection of the intellectual, emotional, and moral content of the individual's life to the ends presented by the working organizations of the society in which he exists.

Developing a system of organized industrial activity capable of surviving under the competitive pressures of technical progress, therefore, is paid for by the increased constraint on the individual's existence. In Freudian terms, men's conduct becomes increasingly 'alienated', 'work for a system they do not control, which operates as an independent power to which individuals must submit'[3]. Such submission is all the more absolute when it is made voluntarily, even enthusiastically.

In the next chapter of this book it is suggested that a social technology, as exhibited in the institutional forms of modern society, has been developed *pari passu* with modern technology in the material sense. Modern organizational forms, governmental and industrial, represent the application of rational thought to social institutions in the same way that technology is the product of the rational manipulation of nature. In the same way, too, it congeals the processes of human affairs—'fixing' them so that they become susceptible to control by large-scale organizations. The reverse aspect of this tendency is the increasing subjection of the individual to the psychological and material domination of the social order, a domination increasingly objective and universal as civilization advances technically.

METHOD

This book is based on all three studies mentioned above. It is a contribution to the study of problems related to the exploitation of scientific discovery by industry in peacetime conditions. It is based on information about what has happened in fifteen firms which have invested in electronics development and five others, four of which had research and development interests in other fields.

We began with an analysis of the situation within each individual firm. But this was made so as to arrive at an understanding of just what is involved in general in relating established industrial procedures to new kinds of activity in which the development of technical innovations plays a continuing part. Our emphasis throughout, therefore, has been on the uniformities or on the range of variety we have observed in this or that aspect of the general problem, and not on the fate of this or that particular firm or department.

The methods of study we have followed are those common to what is called field sociology and to social anthropology. These are simply directed towards gaining acquaintance, through conversation and observation, with the routines of behaviour current in the particular social system being studied, and trying thereafter to reach an appreciation of the codes of conduct which are supposed by the members of the system to underlie behaviour. All this emerged fairly slowly in the course of interviews, meetings, lunch-time conversations, and the like. At the same time we, as outside observers, have tried to construct some systematic explanatory description of what we have been told and have observed, a description which will be reasonably self-consistent and also understandable in terms of other explanatory descriptions of behaviour in other social systems.

All this is very far removed from any method of investigation which could possibly be called scientific. It does not share the principal advantage of anthropological field method, which lies in a lengthy period of residence in the community being studied. Everything has had to depend on what ability we had to appreciate the significance of the things and happenings we saw during our spells inside factories, and to elicit information in interviews and conversations. We had also to learn to distinguish the tones and additives which were occasioned by our roles as outsiders, as academic people, as confidants, as critics.

The centre of interest for us lay in the management itself. If by 'management' is meant a special category of individuals in a concern, it has extremely ill-defined limits, but we have not sought to introduce any more precise connotation. We have rather used the word in its other sense of directing, co-ordinating, and controlling the operations of a working community; and, as we try to make clear, this kind of activity can involve everybody in a concern at different times and in different respects.

Our usual procedure, after the first interview with the head of a firm, was to conduct a series of interviews with as large a number of persons as possible in managerial and supervisory positions. Such interviews lasted anything from one hour to a whole working day. They would start as a general description of the informant's position and function in the concern and of the way in which his job linked with other people's. They would then develop along fairly free lines, taking as their point of departure a request to the informant to be more explicit about some point of significance in his account, which was normally delivered in general, fairly impersonal, terms.

It was during this stage that it proved possible to create a more productive relationship than can be constructed on the basis of one person's seeking information from another. The conventions governing such interviews and the limits of the information regarded as admissible or relevant are nowadays prescribed fairly strictly. To go beyond these limits, it is not enough to demonstrate interest or even sympathy; in the writers' experience, an informant will get to the point of formulating and presenting his experience, beliefs, opinions, anxieties, and criticisms only when there has been established a relationship which is reciprocal in some genuine sense; when there is some point for the informant in going further than the needs of courtesy, and compliance with an undertaking by the firm to co-operate with the researcher, seem to require of him. Thus the researcher has to make the relationship 'real'; one in which he is prepared to behave on his side as what he declares himself to be. This can be done only by showing how he is making use of the information he is receiving; by the occasional interpretation of a situation in terms which are both derived from his perception of the situation as an outsider and as a sociologist or psychologist, and which are also appropriate to his informant's ability or preparedness to comprehend it. From then on, whether the interpretation is accepted or not, there is a freer, more satisfactory quality about the interview, a stronger desire to recruit and present facts, examples, views. There are no interpretations and appraisals contained in any part of this report which have not been communicated at some time or other to persons involved in the situations at issue. Invariably, also, we have found our own ideas being amended, extended, or corrected by such traffic.

After we had become acquainted with the general structure and functioning of the organization, we sought opportunities of observing how people dealt with each other, and also of pursuing, by further interviews, some of the problems of description and interpretation which by this time had appeared. In their simplest and most significant form, these problems were presented as discrepancies between the account of the same functions or parts of the organization given us by different people concerned in them. Such discrepancies, in our experience, are always present, and provide the most direct introduction to the analysis of a situation or social system in sociological terms.

Notes were made of most interviews and formal meetings while they were in progress. It was often possible to record more fully by taperecorder when it seemed likely that this would not prove too disturbing.

ACKNOWLEDGEMENTS

We have been dependent throughout on the co-operation and help of the Scottish Council (Development and Industry) and of the Department of Scientific and Industrial Research and the Ministry of Supply. We have been particularly indebted to Mr. W. S. Robertson, the

Secretary of the Scottish Council, who, while he was Technical Secretary, had executive responsibility for the initiation and conduct of the 'electronics scheme'. Not only did he provide almost all our information concerning the origins of the scheme, but throughout the research and during the writing of the report we have had the very considerable advantage of being able to discuss our experiences and findings with him.

Our largest debt of gratitude is to some three hundred persons who did what they could to provide us with the information we seemed to want and to correct our mistakes afterwards, and especially to the principals of the firms, who made the study possible.

We have also profited by the detailed comments and critical observations on earlier drafts of this book made by Mr. R. G. Stansfield and by a number of senior members of the firms which took part. Responsibility for all statements concerning the firms studied, and for our inferences and comments, remains exclusively ours.

All three studies under review were undertaken by Burns as part of the research programme of the Social Sciences Research Centre of the University of Edinburgh, and were carried out under the authority and with the support of the University's Committee for Research and Co-operation in the Social Sciences. The second study, of Scottish firms, in which G. M. Stalker participated, was supported by a grant from Conditional Aid funds made available by the Human Sciences Committee of the Department of Scientific and Industrial Research. A further grant was made by the Human Sciences Committee in support of the third study, of English firms.

Some of the contents of Chapters 2, 3, 5, and 9 have appeared previously in *Research* (July 1955), *The Accountants' Magazine* (November 1955), *Impact* (September 1956), and *Management Review* (January 1958).

Thanks are due to the Controller of H.M. Stationery Office for permission to quote from *The Administration of War Production* by J. D. Scott and R. Hughes; to Harper & Brothers in respect of *Industrial Sociology* by C. D. Miller and W. H. Form; to Dr. E. L. Leach in respect of *Political Systems of Highland Burma*; to Macmillan & Company in respect of *The Sources of Invention* by J. Jewkes, D. Sawers, and R. Stillerman; and to the University of California Press in respect of *TVA and the Grass Roots* by P. Selznick.

PART ONE

The External Circumstances

CHAPTER 2

The Organization of Innovation

TECHNICAL PROGRESS AND SOCIAL CHANGE

In their most general form, the findings of this research can be put into two statements: Technical progress and organizational development are aspects of one and the same trend in human affairs; and the persons who work to make these processes actual are also their victims.

The connexion between progress in material technology and the emergence of new forms of social organization is familiar enough. But it has become submerged in the century-old controversy about the correct causal sequence of technical progress and social evolution. According to one view, widely held by American social scientists (see e.g.[4],[5]), but stated in its most uncompromising form by Marx, technical progress underlies every kind of change in the social order: 'Assume a particular state of development in the productive faculties of man and you will get a corresponding form of commerce and consumption. Assume particular degrees of development of production, commerce, and consumption, and you will have a corresponding form of social constitution, a corresponding organization of the family, of orders or of classes, in a word, a corresponding civil society.'[6] The argument has an oddly up-to-date ring, not so much in terms of the interpretation of history as of the actualities of international politics. So convinced have we become of the dependence of the total social, political, and economic order on technical development that national output of scientific discoveries and rate of technological advance have begun to appear as an ultimate criterion of culture, and different political and social systems are compared as facilitators of this kind of achievement.

According to the other view, technical progress is the outcome of changes in the institutions of society, even simply, as Durkheim[7] argued, of population growth, which produces not only new needs but the improvements and expansions of knowledge and equipment necessary to satisfy them.

Like many another instance of the chicken-and-egg conundrum, the question which comes first conceals a false antithesis. A social technology, as witnessed in the growth of modern institutions, has been developed alongside technology in the material sense.

The clue lies in Tönnies' perception of the development of modern society as itself a technological process. The relationships and institutions he regarded as characteristic of the modern world are those which enable persons to manipulate others, individually and *en masse*, in the pursuit of their own ends. Thus other people take on more and more of the quality of a natural environment which man looks on as resources for him to consume and manipulate. In many important respects we treat 'human beings like inanimate objects and tools'.[8] It could almost be said that Tönnies regarded social change as one aspect of technical development; organizational techniques and devices for manipulating others were constantly invading the social order.

Progress in power technology, in agriculture, in engineering, in chemicals, and the rest have proceeded—quite inevitably and necessarily—alongside developments in working organizations and in communications, and alongside the elaboration of social and political controls, financial and other economic mechanisms.* Developments on each side have often been by way of adaptation to changes on the other side. Yet there remains the discrepancy, the obvious and awful gap between technical achievement and the constraint and fears which bear upon people in their day-to-day lives. Consciousness of the gap has sometimes, in recent years, found expression in opaque generalizations about man's control of nature out-running his social abilities.

The trouble about such statements is that they mask the realities to which they refer. What is true is that developments in the one have been forced through with just as little regard for ultimate consequences to human welfare as in the other. The advantages looked for, and won, by the progress of technology, both material and social, have been

* Professor Jewkes has suggested that 'Whilst no one would wish to deny that technology and science (in that order) have contributed much to the raising of standards of living in the last two centuries, there is a disposition in these days to exaggerate the contribution they have made and to underestimate that made by new social organisations and institutions.'[9]

immediate or short-run. Very often, these rewards are of the kind which can benefit only individuals or interested minorities. The cost, in more destructive wars, in dust-bowls, in road accidents, on the one hand, and in slums and industrial servitude, emotional deprivation, loss of intellectual and economic independence on the other, only becomes apparent in the long run. It then manifests itself often enough as a price to be paid by others. Unforeseen long-term advantages have of course accrued too. The point is that technological development has typically occurred as a consequence of decisions made in the light of short-term views of the balance of advantage and cost to people in controlling positions.

All novelty involves some degree of risk. The vast majority of biological mutations are said to be harmful. When, as in human affairs, enormous numbers of random possibilities are eliminated by rational choice, the chances of harm rather than good resulting are reduced, not eliminated. The harm consists in both cases in making the individual or organization less fit to survive in its environment than was its predecessor. Very often, the environment of the person or organization is itself changing, so that even to maintain the same degree of fitness for survival, people and institutions may have to change their ways. So the risks attendant upon change may have to be weighed against other risks arising from maintaining the same state of affairs.

This condition of ordinary human existence is made explicit and articulate in the institutions and procedures of industry. And in those sectors of industry in which the creation of innovations is a constant and important part of the total enterprise, the processes of change become visible in an obvious and dramatic way. Here too, the mutual, procreative impact of developments in material technology and social organization finds its clearest expression.

In one very important sense, the link between the two trends is a necessary interdependence. Invention, even more than science, is a social phenomenon; in quite matter-of-fact ways, it is a human activity which can only be fulfilled when certain social conditions obtain, when the inventor inhabits a milieu which prompts him to devote himself to a specific line of work with the promise of rewards—in money, power, or even a secure livelihood, in fame, or even self-esteem—and which will thereafter support him economically and intellectually. The notion of the hermit genius, spinning inventions out of his intellectual and psychic innards, is a nineteenth-century myth, useful then, as myths may always be, but dangerous, as myths always are, once its period of usefulness is past.

If, as Whitehead said, the greatest invention of the nineteenth century was the invention of the method of invention,[10] the task of the succeeding century has been to organize inventiveness. The difference is not in the nature of invention or of inventors, but in the manner in which the context of social institutions is organized for their support.

THE SOCIAL CONTEXT OF INVENTION

The review of the past institutional context of industrial innovation which follows is designed, therefore, to underline the importance of that context and to point out the ways in which it has significantly changed. It is not offered as a history, even in a very abridged form, of the relationship between science and industry during the last two hundred years, but rather as a sketch of the phases of change in institutions of some importance to society. It forms the background to the succeeding account of the attempt by industrial concerns to digest the thing they have swallowed.

During the middle years of the last century the electrical industry was established on the basis, largely, of supplying telegraph services. Within a few years the development of electric motors for tramways and stationary machinery led to very considerable expansion. As new applications multiplied, the need for heavier and more efficient generating plant and distribution equipment accelerated the process. By 1880 there was a flourishing, keenly competitive, electrical industry not only in Britain and in the United States, but also in Germany, France, and other European countries. It was an industry, moreover, in which the technological base was very recent—middle-aged men in the industry would be well aware of the first commercial applications —and in which new applications and design improvements followed each other extremely rapidly. Yet the two major innovations during the last twenty years of the century, incandescent electric lighting[11] and radio,[12] were the work of newcomers, of inventors and enterprises unconnected with the existing industry. No spectacular 'discovery' lighted upon by an individual genius was really responsible; electric lamps and wireless transmission were 'in the air' many years before the first commercial companies were floated.

Swan, a chemist, made experimental incandescent lamps in 1860 which employed the same high-resistance conductor, carbonized paper, as was used in the first commercial lamps marketed twenty years later. There were, by 1880, large industrial concerns manufacturing lighting and other electrical equipment; yet in the event it was Edison who, two years after becoming interested in the possibility, first developed the lamp and formed an independent concern to manufacture it.

Lodge, following upon Hertz's earlier experimental work, demonstrated wireless reception before the British Association in 1894, and two years earlier a physicist had written in the *Fortnightly Review* of the 'possibility of telegraphy without wires, posts, cables, or any of our present costly appliances', adding 'this is no mere dream of a visionary philosopher. All the requirements needed to bring it to within grasp of daily life are well within the possibilities of discovery, and are so reasonable and so clearly in the path of researches which are now being actively prosecuted in every capital of Europe that we may any day expect to hear that they have emerged from the realms of speculation to those of sober fact' (quoted[12]). Yet the development of this obviously profitable venture interested no commercial concerns for ten years.*

In the case of radio, it was the twenty-year-old Marconi who, on the basis of Hertz's work as described in an Italian journal, constructed home-made equipment which was sufficiently advanced after three years' work to communicate messages over eight miles and to bring the Marconi Company into being.

Anyone who has read accounts of technological advances, of inventions, during the nineteenth century will perceive this pattern of development as in many ways entirely typical. It is typical not only of the way in which invention then 'happened', but, even more, of the way people thought of invention as happening at any time. Invention was seen as the product of genius, wayward, uncontrollable, often amateurish; or if not of genius, then of accident and sudden inspiration. As such, it could not be planned for, organized as a part of the field of existing

---

* In America the pattern was oddly repetitive, for in the 1870's Graham Bell had tried to interest telegraph companies in the new, and rival, method of communication by telephone which he had invented; after unavailing efforts he founded the American Telephone and Telegraph Company. By the 1890's this company was the 'most research-minded concern' in the industry. Yet it felt unconcerned about radio (apart from one brief and unsuccessful episode in 1906) until 1911, when the threat from wireless telegraphy was too strong to be any longer ignored.

industry, the idea was intrinsically absurd. In nineteenth-century Britain the archetypal formula for the process of innovation was enshrined in the fantasy of Watt and the kettle.

The fitting of this latter myth to the key episode of the earlier technical revolution was itself characteristic. Of course, the myth of accident and inspiration did go some way towards accounting for the *nineteenth-century* facts. And the outstanding fact was the random distribution of scientific and technical information through the new journals, popular lectures, and societies. These diffusers, and the continued exploitation of major inventions by craftsmen, made it seem possible for any individual innovation to be produced by almost anybody, almost anywhere. Again, the disciplined attack on one difficulty after another, which is how the gap between the scientific idea and the ultimate product is bridged, was still intrinsic to the achievement, but the process was an individual, usually personal, enterprise. Often, as in the case of the electric lamp and radio, many individuals at great removes from each other were involved over a period of years in the development of a single invention.

Images and myths about the past had to fit these contemporary facts. So the boy Watt sat dreaming in front of a boiling kettle and later invented the steam engine. The essential condition of membership of a closely linked group of 'applied scientists', as they would now be called, in the Universities of Glasgow and Edinburgh, the especial circumstance of friendship with Joseph Black, whose discovery of latent heat lay at the bottom of Watt's improvements to the Newcomen engine, the inclusion of the industrialist Roebuck in the circle of personal acquaintanceship—these, the really significant factors, were simply left out of popular account. They were social circumstances which were no longer appropriate to the progress of technology.

Coteries and Clubs

In the latter half of the eighteenth century the Scottish universities were centrally involved not only in the primal discoveries of the industrial revolution in chemistry and engineering but in the technical applications and commercial ventures which exploited them. The rapidity of technological development in so many fields which were being explored simultaneously in the laboratories of Edinburgh and Glasgow was the direct outcome of close personal association between persons with different expertises and different resources. But the association between people like Watt, Black, and Roebuck was founded not so much on their membership of a common profession or organization as on membership of a small, closely integrated society.[13] In the Scotland of the eighteenth century, for such men to be acquainted with each other was virtually inevitable.

Such circles of personal acquaintanceship served as a social medium for a further decade or so. By the beginning of the nineteenth century, fellow-students and friends sought to institutionalize their informal acquaintanceships. Clubs rather than learned societies, as the Lunar Society and the Royal Society of Edinburgh were, they and their offspring and kindred in Manchester and Newcastle, and the archetype in London, included the persons responsible for scientific advance, technical invention and, to a large extent, industrial innovation.

'Towards the close of the last century', says Smiles in his life of Boulton and Watt,[14] 'there were many little clubs or coteries of scientific and literary men established in the provinces, the like of which do not now exist—probably because the communication with the metropolis is so much easier, and because London more than ever absorbs the active intelligence of England, especially in the higher departments of science, art, and literature. The provincial coteries of which we speak were usually centres of the best and most intelligent society of their neighbourhoods and were for the most part distinguished by an active and liberal spirit of inquiry. Leading minds attracted others of like tastes and pursuits and social circles were formed which proved, in many instances, the source of great intellectual activity, as well as enjoyment. At Liverpool, Roscoe and Currie were the centres of one such group; at Warrington Aiken, Enfield, and Priestley of another; at Bristol Dr Beddoes and Humphrey Davy of a third; and at Norwich the Taylors and the Martineaus of a fourth. But perhaps the most distinguished of these provincial societies was that at Birmingham, of which Boulton and Watt were among the most prominent members.

'*The object of the proposed Society was to be at the same time friendly and scientific.* The members were to exchange views with each other on topics relating to literature, arts, and science; each contributing his quota of entertainment and instruction.' (our italics.)

But the rate of expansion of science and technology was too rapid to be accommodated by adapting and multiplying the institutions of sociable intercourse, vigorous as they were in middle-class society at that time. The founding, in 1831, of the British Association, a self-conscious attempt to institute personal links between all scientists and technologists, may be regarded as marking the end of the period when a network of personal relationships on the necessary scale was feasible.

The Diffusion of Technical Information

What took the place of the circle of people who were at once friends, fellow scientists, and business partners, or the coterie whose common interests were at the same time scientific, technological, and financial? Instances of the kind quoted earlier, and the myth, still surviving today, of the lonely inventive genius, suggest that in the period roughly from 1825 to 1875—succeeding the great days of the provincial Societies—information about scientific discoveries became available to a wide variety of people. Personal communication was replaced by mass communication.

The change from speech and letters, involving small-scale contact of a peculiarly intimate and undisturbed kind, to print, by which leisurely and undisturbed communication is effected impersonally, randomly, and with large numbers of individuals, is an institutional change of a particularly potent kind; we are familiar with the part played by printing in accelerating the diffusion of information and ideas in the Renaissance and Reformation, and thus acting as a multiplier of their social impact. A similar function as accelerator and diffuser may be ascribed to the appearance of scientific journalism in the nineteenth century.

There are three phases of this latter-day diaspora, according to the kind of evidence which may be found from the catalogues of journals published by the British Museum. In 1800 there were less than ten scientific journals published in the British Isles. By 1900 there were over a hundred and thirty. The first quarter-century saw the appearance of new general scientific journals, but their numbers are fairly constant (about fifteen) from 1830 onwards. From 1825, the main growth is in specialized and technical journals; there were three in 1825 and over forty by 1860. From mid-century onwards the published transactions and journals of learned societies begin to grow rapidly in numbers, from a dozen in 1850 to about seventy in 1900.[15]

Not only did books and journals appear at an accelerating rate, but clubs and institutes spread the new learning to the utmost limits of literacy in industrial Britain. Attendance at lectures by eminent scientists became as obligatory in the manufacturing towns as was attendance at church and chapel in the country; interest in science, even to the point of patronizing individual scientific workers or building a private laboratory, became gentlemanly. In short, the institutional process which we know as technology—the linking of (a) knowledge of the laboratory demonstrations which established scientific hypotheses with (b) knowledge of manufacturing operations, and of these two with (c) knowledge of existing or presumptive demand for goods and services—this process was spread at random among a very large proportion of the literate population. The fact that it was so spread meant that innovations might appear almost anywhere, might be lighted upon by almost anyone.

The incoherence of the social institutions of technology at the time was reflected in the rudimentary social forms by which innovations were socialized. Technological changes occurred largely through the birth and death of organizations, the simplest form of institutional change. Capital was plentiful, liquid, and diffused, and was readily available for the exploitation of a new device or product. But the institutional build-up around the invention was normally rigid and was identified closely with the line of application and development originally conceived. New concerns had a fairly restricted expectation of life, even if they survived the highly lethal period of early infancy; 'clogs to clogs in three generations' was a piece of proverbial wisdom current in the oldest factory area in the world. The new devices which arose to render the old ones obsolete were generally exploited by new concerns. As Elton Mayo has remarked, the small scale of business enterprises allowed this change to take place without too much dislocation of the social and economic order.[16]

Invention could, and did, make big fortunes in astonishingly short times. The supply of risk capital was relatively enormous. Technical limitations to large lot production were such that the build-up of production had to be slow, and the manufacture of a single device, progressively improved, could absorb all the energies and resources of its designer and backers over a period of years. In this milieu the economic, social, and psychological pressures were all against the organization of research as an industrial resource, and against instituting invention in a professional salaried occupation. The anecdote of Ferranti's weaning from the post he entered at Siemens when he left school reflects the ethos of technical innovation at that time. Ferranti, at the age of seventeen, had invented his first alternator. A meeting with another engineer, Alfred Thompson, led to an introduction to a London barrister:

> 'And you mean to tell me you're content to be at Siemens',' he said, 'earning £1 a week! Good God!'
> Lawyers see life on the seamy side; small wonder then that they become suspicious of all men.
> 'Ferranti', he said, 'if you continue at a job like that, I'll tell you what will happen. As soon as they discover you've got an inventive ability they'll offer you £5 a week and proceed to rob your brains. You'll do the inventing and they'll collect the cash.'
> This was rather bewildering, but it chimed in with certain thoughts that had arisen in Ferranti's own mind.
> 'Perhaps I'd better ask for a rise', he suggested.
> 'For God's sake, don't do anything of the sort', Francis Ince advised. 'Just clear out. That's no place for you. You might stay there till your teeth fall out and never get a dog's chance to doing anything. There's only one thing for you to do. You must start right away on your own.'
> Ferranti objected that he had no capital.
> 'Leave that to me', said his new friend.[17]

But even before Ferranti had his conversation with Ince, the situation was changing. The distribution of scientific information was rapidly becoming organized. The English provincial universities were founded in the second half of the nineteenth century; the major scientific and professional societies were created during the same period.

Professional Scientists and Technologists

By the end of the century, science was the province of groups of specialists working in and supported by universities or quasi-academic institutions. The unity of natural philosophy became separated into departments of chemistry, physics, geology, and later derivatives and hybrids. Information was organized in the form of textbooks and courses; traditions as to what was relevant and irrelevant were created under the authority of qualifying examinations. The intellectual segregation of scientific specialists was promoted by the way in which the new and reformed universities organized studies and teaching. Exchanges of the kind which had been characteristic of the earlier social milieux tended, outside the departmental enclave, to become attenuated and formalized in the meetings and journals of learned societies, where geologists produced papers for other geologists, physicists communed with physicists, and so on. By 1900 scientists were salaried professional men.

On the other hand, the situation of industrial technology was itself changing. When, before the middle of the century, the major scientific discoveries had been and could be the work of gifted amateurs and a few academic scientists, technically competent craftsmen like Maudslay, Nasmyth, and Whitworth had created the machine-tool industry. The engines and machines that were the showpieces of the 1851 Exhibition were largely the work of skilled mechanics and master men who had matched the opportunities presented all around them with the basic training of their apprenticeship, self-acquired mathematics, and a clear grasp of the principles of the new engineering. Yet even then, the development by improvement and new application was becoming a task beyond the capacity of men trained according to traditional craft methods. The outclassing of British products by European competitors at the 1867 Paris Exhibition made this quite explicit. The Royal Commission appointed thereafter to survey technical progress in a number of countries confirmed the impression that Britain had lost, or was losing, the technical lead established in the previous hundred years.

In Britain the answer to the problem was sought in improving and expanding the educational system. It is sought there now; it always is. One may remark at this point the different course followed in Germany. With social distinctions in many social, political, and economic fields more rigid and often more crippling (given the course then set for Western societies) than those prevailing in Britain, yet in one generation Germany overhauled and at many points outdistanced the technical advance of British industry. It was very puzzling.* Perhaps the

* It still is. Sir Charles Snow, in his 1959 Rede Lecture, remarked 'The curious thing was that in Germany, in the 1830's and 1840's, long before serious industrialization had started there, it was possible to get a good university education in applied science, better than anything England or the U.S. could offer for a couple of generations. I don't begin to understand this: it doesn't make social sense: but it was so.'[18]

clue to this sudden acceleration lies in the alliance between the new ethos of nationalism with science and technology, as the other presumptive heirs to the future; the cult of Reason in revolutionary France had set the fashion. The alliance was the orthodox basis of progressive ideas all over Europe, but the arrest of political liberalism in Germany, and its later asphyxiation, may have channelled aspirations and effort much more powerfully in the direction of scientific and technical achievement. Whatever the reason, there is little doubt that the rise of German industry was the consequence of the energy and enthusiasm with which academic scientists like Liebig and the members of the Berlin Physical Society preached their technical gospel and, in the case of the Siemens brothers, themselves created industrial empires.[19] Given this kind of liaison, the appropriate educational system followed. Without it, as in England, the educational system which was devised— in imitation, as it was thought—widened the breach between science and industry.

As the numbers of scientists rose with the foundation of the provincial universities and university colleges, another educational system was devised 'to meet the needs of industry for technical training'. Graduate scientists went for the most part[20] to teach in the schools and universities. For industry, there were the polytechnics and technical colleges, trade schools and evening institutes.

So with the founding of the provincial universities and university colleges, a parallel network of polytechnics, technical colleges, and evening institutes was created. A central examining body for technical subjects was provided in the City and Guilds Institute. By 1902, when local government authorities became largely responsible for all education below university level, the main structure of a separate educational system 'to meet the needs of industry for technical training' was established and lasted for the first half of the present century.

The whole context of industrial innovation had changed. Before 1850 the worlds of science and industry, though separate, had not been distinct; the very existence, on such a large scale, of amateur scientific and technical enquiry demonstrates the ease of access to the world of science enjoyed by anyone with interests which might be satisfied by scientific information. By 1900 science and industry were distinct social systems, entered by different routes, and with very few institutional relationships by which people or information could pass between them. And by 1900, says Cardwell, 'The new applied science industries had left this country, or else had never been started here. In the natural sciences it could hardly be doubted that the lead was Germany's, while in technology the enormous possibilities of the internal combustion engine, for example, were being developed by the French, the Germans, and the Americans. ... Lockyer, writing in 1901, compared our position at the beginning of the new century with what it had been in 1801, at the outset of the railway age—now, the chief London electric railway was American.'[20] (p. 147).

The New Technological System

Eventually, with a continuing need for the gap to be bridged, new social institutions have been developed. The gap became itself a new territory, explored, mapped, and eventually controlled by new specialists, the professional technologists, going by the name of applied scientists or industrial scientists.

Leaving out of account the prior development in Germany of a liaison initiated and purposefully maintained by the scientists themselves, the first successful institution set up to exploit this new territory was not only outside Great Britain, where the worst effects of the separation were experienced, but independent of both industry and established scientific institutions. Edison's Menlo Park Laboratory, employing a hundred workers, was established in 1870. But the entrepreneurial method followed contemporary practice: the concerns to manufacture the new devices were set up and financed as separate ventures.

This earliest model was not followed until the founding of the Department of Scientific and Industrial Research by the British Government in 1917. And until the years immediately before the first World War, very little had been done by industry, apart from chemicals,* to provide the link itself.

Ever since 1918 the development of industrial research in Britain has depended on Government action much more than in the United States.

* The chemical industry had long before this incorporated scientific laboratory work as part of the normal organization of the business concern, but apart from the notable association of Lawes and Gilbert in fertilizer production, the function of the laboratory seems to have been, what it still is in the smaller chemical concerns, to test the product, and control and refine the processes. As in other branches of industry in the nineteenth century, discovery was normally the starting-point of new concerns which exploited it, but firms did not set aside resources of capital and technically qualified people to search for further innovations. In Britain the change came at the end of the century with Brunner and Mond, with the United Alkali Company, and with Nobel, i.e., with the stabilization of the industrial concern.

This may be attributed, as it usually is, to the unenterprising character of British industry in the fields of technical development, to the un-willingness of entrepreneurs to divert resources to development work as being too risky. Yet there were exceptions between 1900 and 1938, notably in chemicals, the industry which had learned from German methods and technological organizations, and enjoyed most stability; and the case is now altered.

It is in the situation of industry as it was in the first decades of the century rather than in such hazy ineluctables as national character that the explanation lies; for such over-caution, such reluctance to take the profit-making opportunities latent in new scientific discoveries, can be explained only by the ignorance of the run of industrialists about the utility of contemporary scientific activity, by the lack of effective means of communication between the two worlds.

By the end of the first World War the need for such communication was publicly acknowledged. Since industry itself was not supplying the intermediary technologists, the Government set up, in 1917, the Department of Scientific and Industrial Research. Between the wars also, the Government supply of intermediary resources increased very considerably with the need to assure the translation of new inventions with military applications into manufactured weapons. It was from these sources that most of the industrial research and development effort in contemporary Britain has grown.

In 1938 Bernal put the amount spent on industrial research, apart from Government expenditure, at £2,000,000, a figure which possibly includes expenditure on routine testing. From a survey carried out by D.S.I.R. in 1955, it was estimated that British firms spent £183 million on research and development during that year.[21]

The Report on Scientific and Engineering Manpower in Great Britain, 1956, put the total numbers of qualified scientists and engineers employed in industry on research and development as 22,000. There is no comparable pre-war figure, but if the figures of scientific staff employed by Defence Ministries and by the Department of Scientific and Industrial Research are taken as a guide, the story is plain enough. Scientific staff were first employed in peacetime by the Army, Navy, and Air Force in the early 1920's. In 1935 the total number employed was less than 500. By 1939 there were 2,000, and by 1951, 15,000. D.S.I.R., which employed less than 300 salaried staff in its research establishments in 1920, had about 1,000 in 1938, & almost 3,500 by 1950.[22]

The work of producing innovations is now largely in the hands of salaried professionals employed in industrial firms, government establishments, or in institutions directly dependent on industry or government for funds.

Technologists (industrial scientists) are now normally people who have graduated in science at a university or a technical college and have thereafter served a period as junior members of a development or design team in an industrial, governmental, or other laboratory. The essential factor is not who employs them, but membership of the appropriate system of communication—electronics, biochemicals, fibres, nucleonics, metallurgy, aeronautics—through which flows information which may contribute to the development of any individual innovation. The comparative independence of technology from industry is reflected in the comparative independence of the technologist. The status of the technologist is a professional one, and there is fairly precise equivalence between ranks as well as salaries in the types of organizations employing technologists. They move freely, and would like to move much more freely, between posts in governmental, university, and industrial establishments. The career is not enclosed by the pressure either of 'loyalty to the firm' or of 'best prospects' within the individual firm.

For the individual firm the technologist is an alien element; he does not fit into the factory system in the same way as other functional specialists, since these are no more than bits of the general management-entrepreneurial function. The actual information held by the technologist, as well as his training and skill, has value outside the firm. This lies at the bottom of the differences in manners, behaviour, dress, and language which so clearly distinguish him from the other members of the firm which employs him.

THE ROLE OF INDUSTRIAL RESEARCH AND DEVELOPMENT

Yet technological progress has become of vital concern for the individual firm in many industries, and the increasing pace of innovation makes it inevitable that the firm provide more and more support for research and development as a condition of its own survival. This is not only because other sectors of industry have become 'infected' by the work of government establishments, or even because industrialists have experienced the profitability of such work, and have overcome their inhibitions about scientific work. There is also a fundamental change in the institutional character of the industrial firm.

The most familiar aspect of this change is in scale, which is a function of alterations in the balance of production and of consumption within the economy. Mass markets have created, and in turn been created by, techniques of mass production; the use of such techniques has made possible certain economies by mere increase in the size of plant.

Secondly, concurrently with increase in scale, there has developed a separation between ownership and control, between the holding of shares and the control of the policy and activities of a company by management itself or by holders of a minority of shares.[23] This tendency is held to be as inherent in the structure of capitalist enterprise as is the tendency towards monopoly in the economy, arising as it does out of the division between the ownership and the use of property.[24] (p. 244)[25] (p. 15). During earlier periods of capitalism economic power resided with the owners of the property, i.e., the shareholders—although in law such owners merely possess documents which give them certain claims against the company, which formally has full ownership. During the present century, however, power has been passing more and more into the hands of the management, of the directors of enterprise. Shares are commonly dispersed among multitudes of small shareholders whose joint influence does not compare with that of a single compact minority interest. The technical and administrative complexities of modern large-scale enterprise have transformed the relationship between the shareholding owner and the manager of productive capacity.

Both these developments have affected the character of the industrial concern. Their influence on the internal organization has been considerable. Increased size has made necessary the division of the general task of management into a multiplicity of individual tasks, each of which has become the province of specialists—salesmen, cost accountants, works managers, designers, planners, secretaries trained in company law, personnel managers, production engineers. Greater administrative complexity, bigger size, and the development of the specialist skills called for have both aided and been promoted by the shift of control from owner to manager.

A significant fraction of resources in Britain have become concentrated and comparatively inelastic. Too much capital, and, more important, too many social commitments are involved in industrial concerns for change to occur through the elementary birth and death cycle usual a hundred years ago. Firms employing many thousands of people cannot close down without wrecking large areas of social organization. Such concerns must keep alive, and in order to keep alive they must become adaptive; change must occur within the organization and not through its extinction and replacement, if it is to occur at all.

Survival of the individual firm becomes a more significant criterion of economic activity the closer the approximation to monopolistic conditions. Keirstead introduces a lengthy exposition of actual pricing policies employed by a 'giant multi-product corporation' as follows: When we define time concretely in terms of the processes of which it is constituted, we are obliged to ask whether the firm aims at a maximum temporal rate of profit ... or whether it aims to obtain the largest estimated profit over some (indefinite) period of time. I am convinced that the latter notion is ... more in accord with real facts of actual situations. ... The maximization of profit at any moment may result in the appearance of competitors whose supplies would reduce price to the point where total net profits over a sufficient period would be reduced below what they might have been had a lower rate been accepted and the potential competitors kept out of the market.'[26]

Directly one introduces time as a function of the profit-maximizing assumption, it is obvious that almost every consideration tends to become subordinate to survival. Directly, that is, the realities of industrial enterprise are organized in terms of the individual firm rather than of the individual entrepreneur, then almost any profit terms upon which the firm can survive become preferable to grosser profits on which it might possibly not survive. There is, in fact, no change in the logical basis, but merely in the way in which it works: (a) through individual mortals, (b) through corporations which are relatively potentially longer-lived. For an individual entrepreneur, profit-taking can be maximized for any period of time however short, since the rewards will certainly be a substantial help towards his own survival. Moreover, for the individual the random sector of circumstances affecting his strategies increases enormously with time. And he makes hay, therefore, when the sun shines, and a bird in the hand is worth two in the bush. For the corporation, randomness does not increase at anything like the same rate. And survival means only survival of the firm. The birds in the bush, which are tomorrow's or the next ten years' production, are just as important as that in the hand.

To sum up: two major changes have occurred in the social circumstances affecting the production of innovations. First, industrial concerns have increased in size: ever greater administrative complexity has brought a wide range of bureaucratic positions and careers into being; control has moved from owners to management. Their survival is therefore a matter of much more intense and widespread concern to themselves and to society; the chances of survival are improved if the technical innovations which might render its processes or products obsolete are developed within it and not by newcomers.

The other change has occurred in the form of institutional relationships and roles within which invention has been possible. The familiar and sociable relationships typical of the eighteenth century provided the ease of communication necessary for the major syntheses of ideas and requirements which introduced the early revolutionary inventions. The scale of scientific and industrial activity rapidly outgrew the social institutions within which the Industrial Revolution was generated; the syntheses which produced inventions and innovations tended to be random or opportunistic. Later in the nineteenth century, new institutional forms introduced barriers between science and industry, and between 'pure' and 'applied' science, as well as between departments of science. In the twentieth century the new and elaborate organization of professional scientists has been eventually matched by one of technical innovators into groups overlapping teaching and research institutions, Government departments and agencies, and industry.

Neither change is complete. Neither set of contrasts is clear. Few sectors of industry, outside chemicals, have fully accepted the changed situation. It was still possible, in the years between the wars, for a major innovation like the gas turbine to be developed in ways reminiscent of the classic days of nineteenth-century back-parlour invention. The jet engine's invention depended on an individual's persistence and enterprise, although the new massive organizations of government and industry were also involved.[27] On the other hand, the career of the most publicized inventor of the nineteenth century, Edison, reflects both the previous epoch, in the almost conscious exploitation of sociable contact with scientists and technologists, and the later, in the maintenance of development groups and the opening of professional careers in invention. Yet the process of change is now far enough advanced for the shape of the forms characteristic of the present system to be discernible.

CHAPTER 3

The Development of the Electronics Industry and the Scottish Council's Scheme

---

The Treasury's *Bulletin for Industry* of March 1958 cited the growth of the electronics industry since the war as the most spectacular success of any section of British industry. Its output was then five times as much as in 1946.

The electronics industry is the technical offspring of the radio industry; indeed, until the end of the 1930's, electronics was more familiarly spoken of as 'wireless techniques'. Electronic products are in essence measuring devices which achieve extreme accuracy and speed by controlling the passage of electrons through a vacuum or other medium. Systems may then be devised whereby a large range of the phenomena which are so measured can be analysed and controlled. The most familiar of these systems are radio and television transmitters and receivers. Before 1939, electronic instruments for use in scientific laboratories were also being industrially produced. Both these sections of the industry have expanded enormously in scope, as well as in volume of production, during the past twenty years. In addition, new ranges of products have been created in electronic computers, which themselves act as the foundation for further systems. By relating electronic devices to servo-mechanisms, another wide range of products, from automatic pilots and gun and missile control to industrial inspection and control equipment, has been added.

RADAR

The transition from 'wireless techniques' to electronics occurred largely through the development of radar before and during World War II.

The history of radar is by now fairly widely known, but a recital of its more prominent features will serve to illustrate, first, the general character of the new relationships between science and industry in Britain; second, the special character of the Government-sponsored form of technological development; and third, the kind of milieu from which electronics development work arose in industry after the war. These are the salient features of the development of radar in Britain:

(i)   The more important facts about radio echoes had been known to scientists for a number of years before 1935. Some four or five years previously Post Office engineers had identified certain interferences with radio transmission in the neighbourhood of Southampton as caused by aircraft in flight.

(ii) In order to dispose of unfounded but persistent claims made about the feasibility of stopping aircraft engines by wireless 'death rays' directed from the ground, the Air Ministry put an enquiry to the head of the Radio Research Station. This elicited the response that the amount of power required for such a device rendered it entirely impracticable; but would the possibility of locating aircraft by radio be of any interest? The Air Ministry, which had been experimenting with sound reflectors as a means of locating enemy aircraft, pronounced itself very interested indeed, and thereafter work started at the Bawdsey Research Station.

(iii) From the beginning of 1935 until 1937 all work on radar devices took place at Bawdsey. In the autumn of 1937 two radio firms were called in to manufacture equipment for the Home Chain of coastal radar stations.

(iv) According to the official historians, 'a final factor in determining the nature of the wartime Telecommunications Research Establishment was the recruitment of University scientists which took place on the outbreak of war. Considered as a move in the organization of scientific effort this was a move of the highest importance. The scientists (mainly physicists) who came to Bawdsey represented a respectable proportion of the younger leaders in their field in the country as a whole. It was certain that their presence at the Telecommunications Research Establishment would attract others. *It was also certain that the work of such a group of men in a field of known* *promise would achieve outstanding success. Similar successes could have been achieved by the same men in other fields,* and the decision to concentrate upon radar, even though it was partly the choice of the individual scientists themselves, was a notable example of the planning of scientific effort.* It was these men, with their research outlook, who, during 1939 and 1940, pressed through the work of evolving centimetric radar.'[28] (Our italics.)

(v) 'The general run of development technicians who remained in the industry developed considerably in their ability to tackle radar problems, but as the scientists in T.R.E. also advanced in technique and in the understanding of production problems they continued up to the end of the war to undertake at least the early stages of the design of the bulk of new equipments and new variants of existing equipments' (*ibid.*, p. 379).

All the essential features of the technological process are contained in this account. There is first the progress of scientific discovery independent of its practical use. Second, there is the direct communication link between user and scientist leading to the definition of (*a*) a need, and (*b*) the scientific information relevant to the design of a device which would meet the need. One may note that the user's requirement eventually met was not what was formulated at the outset but one which was arrived at after some positive interaction; the first request was impossible to meet. Third, the main work of technological innovation was accomplished by men who were trained—and saw themselves—as scientists. Fourth, the creation of innovations by a research and development team was regarded as a natural and indeed inevitable process (see passage in italics in paragraph iv above). Fifth, until the end of the war, the role of industry was that largely of manufacturing articles to new designs worked out in T.R.E. and other Government research establishments. The essential task of innovation could therefore demonstrably be carried on *outside* industry, and be nevertheless effective; it was at least as effective as the design of aircraft,

---

* One of the scientists so recruited comments that what was exemplified was rather what can happen when scientists devote themselves to development from a very strong sense of purpose. Very little planned direction from above was involved in the early formation of the wartime groups of academic scientists; they themselves saw in the development of 'wireless' technology the way in which they could best employ themselves.

which was in the hands of aircraft firms acting on the advice and working to the specifications of the Air Ministry and using the research results of the Royal Aircraft Establishment, the National Physical Laboratory, and the universities[28] (pp. 370–82).

The Working Relationship between Designer and User

The last essential element in the account is the continuity of contact between technologist and user which was so marked a feature of the development of radar. The user's need can never be satisfactorily defined nor the relevant scientific information be made available once for all at the outset of a development project. The crucial circumstance of the first enquiry and the response to it which initiated the work on radar is an archetype of the designer-user relationship. The design which is the outcome is as much determined by the effectiveness of the working relationship between technologist and user as by the individual contributors of either side. This is a matter which will be developed in later chapters. In the present context, it is pointed by the contrast between the wartime organization of radar development in Britain and in Germany.*

In Britain very close contact was established at the outset between the personnel of the Telecommunications Research Establishment and serving officers in the Royal Air Force and officials in the Air Ministry. This system was consolidated at an early stage by the so-called 'Sunday Soviets' instituted by the then Superintendent of T.R.E. A 'Sunday Soviet' was an open meeting held every Sunday in the Superintendent's room to which were invited all senior Air Ministry officials and all Air Staff Officers. The importance of the occasion as a main channel of communication was soon recognized by the Ministry and the Royal Air Force, and every Sunday brought its 'galaxy of everybody from Air Marshals down'. Differences in rank were obscured or ignored. A particular type of equipment or an operational problem would be selected, and the division leaders and group leaders on the T.R.E. side would be there to discuss it. There was thus a very intimate, personal connexion between the people who had the operational knowledge

* This account is substantially that supplied by W. S. Robertson in a personal communication. Mr. Robertson was a member of the T.R.E. wartime staff, and towards the end of the war was engaged on a study of the German organization of radar development and production. We should add that this chapter leans very heavily throughout on information supplied at various times by Mr. Robertson.

and the problems to face, who saw their men getting shot down, and who themselves flew and got shot down, on the one hand, and on the other, the people who worked in T.R.E. and who had an intimate knowledge of the scientific techniques and their scope. The result of this was that the laboratory workers got an immediate emotional as well as intellectual appreciation of the pressing operational difficulties, needs, and problems which they could not have acquired by any other means. Equally important, the operational people began to acquire notions of the potentialities of the techniques which they could never have had except through meeting the people who had originated them. The rapid application of the techniques to the problems which did take place was made possible by this intimate joining of operational needs with technical possibilities in an immediate, personal, informal way.

In Germany, a great deal of radar development had taken place before the outbreak of war, and some equipment of an advanced type was in quantity production by 1939, but work on development was virtually closed down for a time because it was thought there would be no further need for it. When the Germans did start up again, they established a Plenipotentiary for High Frequency Techniques. This official established a chain of new research institutions. He also introduced a system of logging all the available laboratory effort not only in his own institutions but in all the industrial firms, universities, and technical colleges in the country. He then established contact with an official in the Air Ministry corresponding to himself. The Air Ministry official defined specifications of what the Air Force wanted, and these went to the Plenipotentiary for High Frequency Techniques. The latter would then consult his list and see which laboratories were unemployed, and then post off the specification to one of them. The laboratory would thereupon make an equipment designed without any real knowledge of the operational needs and therefore, in many cases, not meeting them adequately. But much more important than these deficiencies was the fact that most of the possibilities were not realized anyway, because the operational people could not envisage the potentialities of the techniques available, nor could the technical people appreciate the problems of the men who were flying machines.

There is no reason to believe that the Germans were technically inferior either in research or in production methods at the outset of the war, or that they were lacking in resources. The clue to the strong lead which Britain obtained and held seems to lie in the appreciation of the supreme importance of bringing technical knowledge bodily into close contact with the user's requirements.

THE POST-WAR SITUATION

At the end of the war, the electronics industry consisted of a number of small new firms, and of departments of larger, older firms which had grown up and become sizeable ventures behind barriers or at a distance which sealed them off from their parent companies. Some development work for the Government might still be needed in peacetime, it was thought, but production of military equipment virtually ceased. There was a market for radar techniques in air and marine navigational aids, and some firms recruited research and development engineers in anticipation of a steep growth in the manufacture of civil aircraft in Britain. Telecommunications had made very big advances during the war, and was on the threshold of further revolutionary changes. Television would provide a new mass market in addition to the replacement of millions of out-of-date domestic radio receivers. The assurance that a much bigger proportion of national resources than before the war would be devoted to science and to technological development persuaded a number of firms to put money into the development of new electronic instruments.

The rapid growth of the industry, however, is based on something more than these reasonably safe bets. During the war, the section of industry which had been implicated in radar work had undergone a kind of metabolic change. By force of circumstances, a number of firms had not only witnessed but themselves experienced time and time again the successful and profitable application of scientific information of an abstruse and 'etherealized' kind, to use Toynbee's word, to the design and manufacture of new devices. Some of the larger firms, especially valve manufacturers and those in the television race, had seen a little of this immediately before the war, but such work had been the exception; industrial research in electrical engineering was normally devoted to the improvement of systems stabilized on fairly conventional lines. What was new in 1945 was the feeling—still perhaps something short of confidence—that an enormous, possibly limitless, field of enterprise could be opened up by harnessing technical information to practical and commercially profitable uses along lines similar to those which had paid so handsomely in the military field.

Industrialists had come to see research and development, and industrial scientists, in an entirely new light. Not only had the communication gap been bridged, the prestige of science and scientists had grown; the reflection of this prestige on the managements which employed such people, the sense of élite status which derived from membership of the new technological club, played no small part in the decision of many concerns to invest in development laboratories.*

This new vision might indeed have been short-lived had it not made its appearance at a time when, for as long into the future as anyone cared to look, the economic situation of Britain and of the world would be dominated by scarcity rather than by surplus, as it had been for as far back as anyone cared to remember. Other circumstances: the austerities of the post-war years, which built up reserves and kept distributed profits discreetly low, the difficulty of investing in buildings and equipment, the release of scientists and engineers from war work; all combined to urge concerns in the direction of building up their own research and development resources.

At a critical moment, a defence programme was launched which called once again for a large industrial contribution to the development of new weapons and military equipment. For those firms which were in a position to benefit, therefore, research and development organizations could be expanded in a very economical fashion.

POST-WAR GROWTH OF THE INDUSTRY

The eight English concerns which took part in the study represent a cross-section of the industry which has grown from these beginnings. Their product ranges included domestic radio and television receivers; telecommunications and exchange systems; laboratory instruments; radar equipment for defence, for aircraft and shipping navigation, and for meteorology; analogue and digital computers; industrial control systems; semi-conductors; machine-tool control; broadcasting equip-

* Anticipating some later discussion (pp. 112–19), we may interpret what happened in this sector of the electrical industry as a significant extension of the boundaries of the feasible. In terms of Shackle's model of business expectation, the 'degree of belief, measured inversely as potential surprise' in a whole range of possible ventures was considerably increased, with a consequent extended climb of φ values up the contours of value representing the entrepreneur's interest.[29] Put in another way, the 'aspiration level' of these entrepreneurs was raised.[30] As we shall also see in Chapter 4, Simon's thesis that opportunity cost may be read as aspiration level was also empirically true for a number of new entrants to the industry after the war (ibid, p. 55).

ment (radio and television); guided weapons. Many of the firms visited had many other product lines, but these were the responsibility of other parts of the firms, and clearly distinct from those concerned in the survey.

All eight English concerns have grown very rapidly since the war. One of the largest had in 1958 sixteen times the capital and number of employees it had twelve years earlier. Another, formed in 1950 with a staff of 150, employed 3,000 by mid-1958. A third had almost doubled its turnover in each year since 1951, when it first took on its present shape.

There are six aspects of this expansion: first, there is the 'natural' expansion of demand in, e.g., the home television receiver market, defence work (until recently), broadcasting, telecommunications, navigational aids. Some firms have, in addition, increased their slice of such markets at home and abroad by seizing the opportunities provided by a technical lead, by superior market intelligence, or by crucial economies in production time or cost.

Second, 'horizontal' moves have been made into fields related to the existing products; e.g., from radar into nucleonic instrumentation and into display equipment, from fire-control into machine-tool control, from radio receivers into gramophones, from instruments into industrial control and data processing systems.

Third, a few vertical expansions have occurred through the need to control supplies of critical importance to major products. Manufacturers of television and radio receivers, particularly, have acquired interests in cabinet making and plastics. In engineering terms, however, there has been hardly any inclination to take over the manufacture of components. Vertical expansion has normally accompanied general expansion and is merely an aspect of the growth of machine shops, sheet-metal shops, etc.

Fourth, new product ranges have arisen from the application to known user requirements of techniques learned in the course of development work of another kind (usually for Defence Ministries). The most spectacular of these is in data-processing and the application of computers to the continuous guidance of machine tool operations.

Fifth, new product ranges have developed out of systematic study of potential user needs in relation to the commercial aims of the firm and its resources and technical abilities (e.g. radar navigation, computers, business machines, display techniques, industrial controls).

Sixth, new ventures have been initiated in association with other British or with foreign companies to facilitate entry into new markets, or to utilize expensive resources (e.g. an international sales organization) more fully, or to cement an association already profitable in other terms by joint exploration of new techniques (e.g. colour television).

THE SCOTTISH COUNCIL'S ELECTRONICS SCHEME

The part played by the factors which we have enumerated in the creation and growth of the electronics industry may be made somewhat plainer by the course of events in Scotland, where a deliberate, planned attempt was made to bring an electronics industry into being. In particular, two circumstances have been of critical importance: the role of the Government as the chief customer, and membership of the 'electronics club'.

For Scotland, the pertinence of the new technology to industry was sharpened by the disquiet which persisted as a legacy of the depression of the 1930s. During the eighteenth and nineteenth centuries Scotland had taken a leading part in new engineering and chemical developments. A leading position was still maintained in later years in marine engineering, boiler manufacture, locomotive building, and structural engineering, but that leading position for many of the firms concerned was due to their scale as producers rather than to their lead in the introduction of new techniques or products. And in the important new industries in engineering and other sectors which had grown to a dominant position during the first half of the century—electrical engineering, photography, motor-cars, aircraft, rayon, wireless, telephones, plastics, pharmaceutical chemicals, non-ferrous metals and alloys, fuels—Scotland had in 1945 little or no share. The advent of a new crop of industries based on new techniques meant the likelihood of the gap between the technical base of Scottish industry and that of industry in England and overseas being disastrously widened.

The consciousness of the weakness latent in Scottish industry behind the high level of employment and the full order books was largely responsible for the foundation of the Scottish Council in 1947. From the start, it was committed to a policy of encouraging not only industrial expansion in general, but the growth of kinds of enterprise new to Scotland. As a first step, and because it was a field familiar to the people

45

centrally concerned, it was decided to formulate a planned attempt to promote the electronics industry in Scotland.

The first step towards the electronics scheme was taken when the Scottish Economic Conference of July 1948 approved two documents prepared by the Research Committee of the Scottish Council, which argued for an increased allocation to Scotland of research and development contracts placed by government departments (especially the defence ministries), for an expansion of the provision made for research and development within individual firms, and for better liaison between industrial development establishments and universities and other sources of scientific information.

The next stage was a meeting in Glasgow just before Christmas 1948 of officials of the Ministry of Supply with one or two heads of departments at Glasgow University and the Royal Technical College and representatives of the Scottish Council. The meeting consisted of a series of appraisals of the existing possibilities for industrial development in a number of technical fields, and of the teaching and research plans of the universities. From all this information, it was clear that a good deal depended on the Government's consenting to place more development contracts in Scotland—on repeating, on a small scale, the war-time expedient of deliberately cultivating new industrial potential. On the other hand, with the post-war growth of electronics work in the Scottish universities, Scotland would be excellently placed so far as recruitment to development teams and source laboratories was concerned.

Beyond these matters, further definition was given to the strategy of the eventual electronics scheme. In reviewing the relationship between research sources and industrial development, Sir Ben Lockspeiser pointed out that at the beginning of the war the research establishments had been overloaded with technical problems of a straightforward mechanical engineering nature. Early development contracts had been designed to move as many as possible of these problems out into industry. It was quite essential that in any build-up of development work in Scotland, where such work was almost entirely novel ground, some machinery should be provided by which technical information—often of a mechanical engineering or other nature, only indirectly concerned with the system under development but all-important in designing for production—should be obtainable from an intermediate source between the firms and the government research establishments. If such a staging arrangement were devised, the distance of Scottish firms from the government research establishments was not so strong a disadvantage. All this meant, he concluded, that there would probably have to be some departure from the methods hitherto used of the responsible officer in a government establishment choosing the individual firm which might be suitable for a particular development task. Instead of a centralized system for allocating tasks and for building up new industrial potential, a number of firms in association might operate an intermediate establishment which could incorporate research resources ancillary to the Government establishments and also negotiate for development contracts on behalf of the group.

In order to obtain official sanction for the effective pursuit of the suggestions put forward at this informal Glasgow meeting, it was followed fairly quickly by a formal meeting between representatives of the Scottish Council and senior officials of the Ministry of Supply. The Scottish Council advanced the case made earlier at the Scottish Economic Conference, with some modifications of emphasis. Very briefly, that case was to the effect that the Government's policy, as defined in its Distribution of Industry Act, was to ensure a more even development of industry throughout the industrial areas of the United Kingdom. To this end it was spending large sums of money in the Development Areas on industrial estates and services. So far as explicit policy was concerned, that is, the Board of Trade was regarded as entirely responsible for exercising what control or influence the Government had on the location of industry. The Scottish Council argued that so far as some of the most important industries for the future were concerned—electronics, precision engineering, aircraft production, atomic energy applications, and one or two others—the decisively important influence was that which was indirectly exerted by the Ministry of Supply in placing contracts for development work, and not the explicit powers of the Board of Trade. It was therefore unreasonable that the Government should on the one hand spend large sums of money on encouraging the growth of industry in Scotland, while on the other it should operate machinery which inhibited the growth in Scotland of the most important sections of industry.

This submission, elaborated in discussion, was again well received, and the Ministry of Supply agreed to co-operate with the Scottish Council in furthering development work in Scotland in those sectors of industry with which they were mainly concerned.

Hitherto, discussion had proceeded in a general context of Service requirements very different from that which had obtained during the war and later. In 1948 and for most of 1949 thinking was in terms of a fairly low level of armament requirements. This meant that defence ministries were under no special pressure to build industrial potential to a higher level than they were left with after the end of the war, so that when the discussions of this specific and practical kind began, it soon became clear that a vicious circle existed. On the one hand, the Ministries were only too willing to place more research and development work in Scotland than they had done in the past, if the Scottish Council could show them where to put it. But on the other hand they were looking for concerns with an efficient development team with at least six or seven design engineers, the larger the team the better. There were already any number of smaller firms in the south who were clamouring for work; the Government had far more capacity of that kind than was needed. In general the officials concerned could not envisage the effort being generated for technical and administrative liaison with an appreciable number of such small firms at such a distance. So, at that time, the position appeared to be that since there was no development capacity of the size the Ministry wanted, apart from Ferranti Ltd. in Edinburgh, no contracts could be forthcoming; because no contracts came, firms could not acquire the technical staff, the technical knowledge, and the initial financial assurances on which to expand their electronic work.

The only escape from this impasse seemed to be the staging arrangement suggested at the early Glasgow meeting. This suggestion had no specific reference to electronics; it was, in fact, apropos of experience in another kind of development work. It was moreover directed to the solving of problems relating to the breakdown of development work and not to this fundamental question of how to originate development potential. Yet some such scheme might be applicable to the commercial organization of development contracts. The Scottish Council decided to explore the possibility of forming a co-operative group of firms which would, in combination, offer the prospect of a single organization large enough to attract Ministry contracts.

The next phase, therefore, was that of testing out the reaction of Scottish firms to the idea, necessarily only half-formed, of a co-operative venture into the field—entirely new to them—of electronics development. In one respect the aim was quite precise. There was no intention of building up firms to supply the domestic radio and tele-vision market, nor of involving firms for whom defence contracts would be a substantial part of their development and production effort in the long run. The objective was then, as it remained, the intrusion of electronics technology into the engineering industry so as to create new applications—most of them as yet unrealized—which would become visible to specialist engineering firms as they became familiar with the new techniques.

A formal paper was drawn up embodying a scheme of co-operation, again in the form of a submission to the Secretary of State for Scotland. Ferranti Ltd. was to act as a centre, negotiating research and develop-ment contracts larger than it could itself fulfil but suitable for breaking down into parts, normally two parts, one of which would be manage-able by a firm new to the work. Ferranti Ltd. would be responsible for part of each contract, and would also perform intermediary functions, undertaking the general responsibility for completion of contracts and acting as a local source of technical and other information. This latter function would include taking technologists recruited by the other firms into its own laboratories for training in development and design work. Training would take the form of working with the Ferranti development group responsible for their part of the contract for a period sufficient for the new development team to become acquainted with requirements, methods, and the specific nature of the design pro-blems with which they were faced.

Nothing happened thereafter to alter the main outlines and objectives of the scheme, but a great deal of time was consumed in steering it through various vicissitudes, the principal being the decision of the Government, at the end of 1951, to reduce expenditure on all new building. This delayed work on the new central laboratories which were to have development teams beginning work on defence projects. One of two firms began work on development contracts, but in general the scheme had to await the completion of the new laboratory in October, 1954, before it could operate as planned.

The scheme worked in the following fashion. First contact was usually between a member of the staff of the Scottish Council and the manage-ment of a particular firm; either the Scottish Council considered the firm to be potentially or actually interested in electronics, and to have the right resources, especially in machines and draughtsmen, and sought a meeting with the management; or the latter became interested of its own accord and approached the Scottish Council. At an early stage, in any case, the firm was informed of the opportunities offered by the scheme and of the procedures which might be regarded as normal. Thereafter, the management was invited to look around the central laboratory at Ferranti's; in many instances it went further and paid a visit to one or two of the Government's research establishments and other organizations which might be involved in development work appropriate to their concern. If the firm retained its interest after this exposure, there was then a time of further exploration and discussion between its management, Ferranti Ltd., and the Scottish Council. There would also be discussions with one of the defence ministries about a development contract suitable to their present activities and level of technical proficiency, and to their own plans for the future. A suitable development contract was one which could be handled by the firm *and* Ferranti Ltd. between them. When this was signed, or about then, the new firm engaged a number of design engineers who then worked alongside the Ferranti group attached to the contract. When the new firm's development group was considered, and felt itself, to have acquired sufficient grasp of the requirements, of the special techniques involved, and of the method of working, it moved to its own premises, where a laboratory had meanwhile been prepared. Thereafter the contract was carried to its conclusion, close liaison with the Ferranti group being maintained throughout.

Not surprisingly, main contracts of this kind proved infrequent, and Ferranti Ltd. played a much more positive role in the formative years by arranging (with Ministry approval) for new entrants to take on parts of their own development projects on a sub-contract basis.

After this beginning, the firm was expected to acquire development contracts on its own account by direct negotiation, and each firm naturally hoped to be well placed for production contracts—much more profitable than development contracts—which might accrue from development work successfully accomplished. But up to this stage, and even a little beyond it, was all in the nature of investment for the future. For no firm with reasonably active production lines of its own would be attracted into work which offered such a low return as do development contracts, or in which the biggest incentives were the promise of defence contracts, which are a notoriously unreliable foundation for a manufacturing concern to build on. Often a design is superseded before it is completed: production contracts tend to be fairly infrequent nowadays, and fairly small when they arise. The principal attraction to the newcomer (and the purpose of the scheme) was the possibility of applying the knowledge, skills, and organization built up through these initial defence development contracts to the improvement of existing processes and products directed towards the commercial market.

The role of Ferranti Ltd. was not entirely that of fairy godmother. The laboratory building served as the central establishment of the scheme, not only for training purposes but also as something of a source and exchange centre of ideas and information in Scotland—a kind of sub-station of the Government research establishments themselves; but the main users of the building were undoubtedly Ferranti Ltd., alongside whose factory it had been built. More than this, however, Ferranti Ltd. hoped to profit, along with other member firms, from the pooling and exchange of ideas concerning mechanical design, engineering, and production methods, all of which enter very largely into successful development work. The field of potential expansion for electronics was so vast, and the threat of American and other competition so considerable, that the interests of any single firm heavily committed to the electronics industry were seen as best served by any acceleration of development in Britain that could be contrived; for the industry, it was said, is essentially national in structure, requiring as it does to be rooted in an extensive field of 'pure' and 'basic' research in the universities and other research establishments. Success for the individual British firm, even the large concern, was much dependent on the growth of 'electronic-mindedness' in all sectors of industry.

Thus the scheme itself aimed at being no more than a pump-priming operation. The initial share of a development contract could help a firm recruit a design team. By working alongside an experienced team in Ferranti's the team could gain invaluable experience in appreciating the requirements, the special techniques and appropriate methods of working together. But thereafter the firm was expected to find its own feet, and not only to take on development contracts alone but to exploit the potentialities of electronic techniques in its own particular sector of industry and in its own markets. It was the attainment of this kind of fitness which was the real purpose of the scheme, and the real attraction to firms entering it. And this last phase was the focal point of our own study of the firms which entered the scheme.

CHAPTER 4

The Market Context

So far we have been concerned with some very general observations about the character and antecedents of the world inhabited by the firms which participated in the studies. Turning now to the activities of the firms themselves, the undefined and passive notion of 'meeting the conditions set by the world around it' becomes the more familiar 'ways in which the firm achieves its purpose'. This purpose, which is the overt and expressed intention of the firm, as against any latent, implicit function, is to survive and grow—to achieve commercial success.

UNCERTAINTY AND THE QUEST FOR MARKETS IN THE POST-WAR PERIOD

The years immediately following the war, when awareness of its own existence dawned on the new industry, when it found itself in possession of enormous resources of technical information and no customers, were as critically important commercially as the war years had been technically. The pace of development, the system of management, sales techniques and policies, levels of investment, all had to be established. These are matters which are ordinarily worked out as an industry grows and on the basis of experience in the parent or allied industry. In electronics, the industry had grown almost to maturity before these requirements made themselves felt; and both conditions and norms were without useful precedent.

The kind of situation confronting the individual firm has been described in some detail by the manager of the Edinburgh factory of Ferranti Ltd.*

'In May, 1942, Ferranti Ltd. were asked to undertake the engineering design of a new gyroscopic gun sight. . . . The basic experiments and design had been carried out at the Royal Aircraft Establishment, Farnborough, but the design had to be modified for production and the necessary drawings and samples produced.

'In October, 1942, Ferranti Ltd. were asked if they could undertake quantity production immediately on completion of development. At this time we had 13,000 workers engaged on war production in Lancashire and had no spare capacity available. However, the Ministry of Aircraft Production impressed on us the importance of this new equipment, as it was expected that the percentage of aircraft destroyed would be greatly increased, and asked us to consider sending a team to Scotland to undertake the work.

'Technical staff, production engineers, and foremen who had experience of this small and intricate type of production were transferred from our parent factories in Lancashire, 40 men were taken from the Ministry of Labour Training Centres and sent down to Lancashire for two months' training as machine setters. The rest of the labour, up to a peak of approximately 1,000 people, was recruited locally.

'Despite these problems the first gun sights were produced by the end of November, 1943—in advance of the Americans, who had adopted this British design at the same time as ourselves—reaching a peak production of 1,000 per month, and supplying the bulk of the Royal Air Force requirements.

'It is important at this stage to understand that Ferranti Ltd. is made up of a group of Departments, each dealing with a different aspect of the electrical industry, and each having its own Management, Design and Works Organization. Although we had been given, and there is always available, considerable help by our parent organization, we are free to develop our own ideas, but we must make sure that our successes are not exceeded by our mistakes; in other words we must make a profit.

* We are grateful to Mr. J. N. Toothill and to Professor R. S. Edwards for permission to quote from Mr. Toothill's paper 'Starting a New Industry in Scotland' presented to a seminar held at the London School of Economics in the series on Problems in Industrial Administration, 1951–52 (Paper No. 15).

'At the end of the European war we began to take stock of ourselves to find out what we knew and how we could turn this knowledge into products. We had built up a good works organization on fairly accurate instrument production. We had a small design staff and drawing office, but this was much too weak and too highly specialized—we were a body with a tiny head and we had no future unless we could design equipment for which there was a demand.

'Our first reaction was a general awareness of the growing use of electronics, both for control purposes and for radar, and it was obvious that there would be a market somewhere for a combination of precision instruments and electronic engineering.

'We had just started to grow our design head when V.J. day was announced. This meant cutting down our wartime output to 10 per cent, and we began to worry about keeping together some of the skilled effort it had been such a hard job building up. We took on the manufacture of equipment for other people, but this proved a big mistake. We know how one normally under-estimated initial production difficulties of a new product, but even with this knowledge we still did so.

'However, we were . . . able to attract one of our senior electronic engineers from the Ferranti organization in Manchester, and we started up a design team on electronic devices. This was followed in 1947 by a laboratory to take over the radar interests of the Company, and a little later we welcomed the transfer from Manchester of the Company's physicists working on specialized radio valves.

'As soon as we had decided to build up a design team, we gave serious thought to the markets which would use the techniques that we were building up, in particular the aviation market, i.e.

'(1) During the war we had carried out a little engineering and manufactured a small proportion of a military aeroplane. We decided therefore to take on a nucleus of engineers from the Government establishments in order that we should have a sound knowledge of operational problems, and so that we could educate our existing and new design staff.

'We wanted to be in a position to take on a defence problem from the initial requirements through flight trial stages to final production, instead of only engineering and production.

'(2) We also noted a trend in military aviation to more complexity in instruments and electronics, and thought that this would increase rather than decrease because of the higher speeds and operating heights of jet aircraft.

'(3) Furthermore, if we could secure Government design contracts, we should be able to cover a wider field of engineering and scientific endeavour, as well as establishing contacts on a more extensive scale.

'We came to the conclusion that the Services would want to design equipment and, so to speak, keep their technical head above water, but they would not be placing much in the way of production contracts. This, like most of our predictions, proved only moderately accurate, some situations being slow to develop but in other cases the position changing much quicker than one expected.

'(4) Civil aviation was also a possible market. We had been brought up on the knowledge that the growth of sales of electricity was exponential at a rate of 7 per cent cumulative per year. We knew that the size of the British air lines of that time was very small, but we plotted a growth curve from 1921 and found that the number of passenger miles flown on British air lines was increasing at a cumulative rate of over 20 per cent per year. We weighted the numbers of aircraft in the three fleets by factors to allow for increased utilization, and found that the trend was to fewer but faster and heavier aircraft for the same number of passenger miles flown. All this added up to in our minds was that it would be a difficult business until after 1960, and then it might be a worth-while market, but that if we wanted to be in it in 1960 the time to start was 1946.

'We had started our radar team with a nucleus of engineers from the Government establishments, and partly because of this we were entrusted with the design of the radar distance measuring equipment planned as an international civil navigation aid.

'(5) Besides looking at the military and civil aviation market, we realized that the country's full employment policy would call for an increase in mechanization, and that there might be a market for a combination of instruments and electronics for this purpose. It is quite a difficult job to design this type of equipment to improve the standard engineering machine-tools because of the variety of work that has to be carried out on each type of machine, so that we turned our attention to process industries, particularly textiles. We designed a certain amount of equipment after talking to friends in particular companies, but we were never able to find unanimity of opinion on what was required. We reluctantly came to the conclusion that we did not know enough about process industries to back our own judgement on a particular line of development. Recently we have turned our minds again to our own manufacturing problems, and we feel we have the design knowledge to make slow but substantial progress in this field.'

The episodes of this narrative neatly itemize the various genetic factors discussed in the first two chapters: the origins of an industrial enterprise in the work of a government research establishment; management's apprenticeship in the industrial organization of scientific effort; the faith in the future of technological development as an industrial activity; the recruitment of a laboratory staff as the foundation of the concern's future; the widening prospect of electronic applications in the early 1950s, with the final hint of the work then beginning on the Ferranti computer-controlled machine-tool.

The Government Market

But the account serves as an introduction to this chapter because of the light it sheds on the way the management assessed commercial possibilities at an all-important moment in the career of the enterprise. There is a striking difference between the confidence with which the firm approached the familiar market of military and civil aviation— both of which were Government-controlled and had developed elaborate and systematic methods of defining their requirements—and the tentativeness and lack of information which appear from the reference to the experiments with the process industries and other sections of the ordinary commercial market. These latter were, it is true, relatively unfamiliar—although a firm established for two generations in Manchester could not have been wholly unacquainted with textile industries—but the strategy implied in the account reflects a view of a market, even a domestic market, as a trackless, mysterious, and dangerous jungle to be explored on foot and with the aid of friendly headmen rather than surveyed and mapped from the air. In the event, the main line followed for the build-up of business to compensate for the ninety per cent drop in government contract work was that of expanding in a familiar market, even though this was, in the case of military aircraft, expected to decline. To do this, the firm imported radar development engineers and built up a new radar development group. Familiarity of customers was more important in determining the direction of development than knowledge of the potentialities of new techniques, although this too was of course present.

With all the advantages of hindsight and of the later experience of firms in the industry, it is now possible to see that there was, beyond the obvious difference between working on government contracts and selling ideas and products commercially, a latent difference which affected the attitude of the firm to the whole task of selling its work in an open market. This difference is one which persisted for the next twelve years, during which the Government was the principal customer of the industry.

Defence ministries are customers who define their needs with clarity and much detail. They are in constant and intimate touch with research establishments, which inform them of the potentialities of techniques. They are actively concerned to keep industrial development and production alive and well. The research establishments themselves have hitherto been the largest, and a quite essential, source and reservoir of technical information. In a very real sense, the market has nursed the industry which served it. In an important sense, the customer has provided, or reduced the need for, a sales service.

The compulsory exodus from these sheltered market conditions at the end of the war was a risky and uncomfortable business for most firms; so much so that there was very considerable relief when in 1949 the Government announced the new defence programme. Most firms thereafter reverted to a policy of heavy dependence on Government contracts.

However, during the 1950s, as the industry grew towards maturity, disadvantages began to make themselves felt. As the managing director of one of the largest English concerns put it, 'At the end of the war, we naturally asked the Ministry of Supply whether they would continue to want us. They said no. Two years later, as you will appreciate, they were after us again, wanting us to build up as fast as we possibly could, and from then on nine-tenths of our work was for the Ministry. We didn't regret this. We don't regret it now, because it gave us very valuable background knowledge. But it was, for a business concern, very dangerous to depend so very much on Ministry work, because there was always the possibility of withdrawal of contracts, of policy changes which would affect us very seriously.'

The dangerous subjection of the 'government market' to the vicissitudes of national and international politics was never very far from the minds of people in the industry, and their fears were realized by the 1957 Defence White Paper. However, the 'defence cuts' only accelerated or anticipated a shift towards commercial markets that was already on the way, partly because of the 'natural' expansion of demand for electronics equipment by industrial and business concerns and by foreign governments, partly because of the disinclination of some firms to remain dependent on Government work.

The main reason for this change of heart was apprehension about the future volume of government development contracts, but in addition such work showed low profit margins without there being any assurance that production contracts would follow. More than one firm mentioned production contracts following on successful development work which were snapped up by other concerns which—probably because of spare production capacity—quoted 'absurdly low' prices. It was also said that personnel changes in defence departments made for difficulties and back-tracking in development progress. One managing director thought that there were occasions when technical officers might experience some need to justify their positions by needless elaboration or alteration of a specification. Two others thought that there had been a general decline in the technical expertise available to defence ministries (i.e. in the research establishments).

While every electronics firm, at the time of the studies, had a good deal of government work on hand, some regarded these disadvantages as overwhelming. Indeed, two of the firms largely engaged in domestic radio manufacture showed no inclination to regard government development contracts as worth while in financial terms. In both, development groups on government contracts tended to work rather in isolation from other development and design teams; both firms saw government work as not only unprofitable, but also as virtually sterile so far as technical ideas for commercial application were concerned. In general, they kept some government work on as a form of insurance against a shutdown which a 'national emergency' might otherwise force on them. There were minor assets: it attracted good engineers, ideas on circuit design, etc., *might* appear from it, it *might* serve as a source of new products and a nucleus for growth in fresh directions if the main market became very difficult—though even here, private development work was obviously much more serviceable.

There was a very marked contrast in one radio manufacturing

concern between the isolation of the government development contracts group, and a new development laboratory devoted to work of clearly commercial application. The main contribution of the first group admitted by the technical director, after some brain-searching, was 'some work on varying climatic conditions—but I expect not much'. The other laboratory, supported entirely by the firm, provided wider technical interests in much more realistic terms, together with the prospect of direct commercial exploitation of its own work; the link was most clearly visible in, for example, the close personal contact between a group in the radio design laboratory concerned with the use of transistors in the circuit design of domestic sets and the semi-conductor section of the new laboratory.

By 1958, 'government contract development' groups, on which the industry had been founded, were being regarded as the 'poor relations' of the main development and design groups in four out of eight of the English concerns. The later 1950's, therefore, have seen a recurrence, in milder form, of the situation of 1946, when the industry was faced with a fundamentally different market situation. This time the industry was much better fitted, technically, to deal with the change, and customers were much more in evidence (as were foreign competitors). But the change proved in many ways no less difficult. Government work in electronics development—and production—was founded on the closest possible relationship between designer and customer, a relationship in which the 'sales function', in the sense of a third, intermediary, job of canvassing and negotiating, was largely superfluous. Such functions as market exploration, market development, the study of trends, and the continuous adaptation of the firm to emerging needs had been either written into the user task or discharged in an implicit way largely by the technical people concerned in research, development, and, to some extent, even production both on the Government and the industrial side.

The Commercial Market

The fact that the sales relationship in the government market was in large measure initiated and maintained by the user, and reinforced by his dominant technical role, educated most firms into unfitness for commercial markets. For many concerns have had in latter years not only to create organizations and to find people to undertake these fundamental sales tasks in the new commercial markets, but to discover that such tasks exist.

Not only are financial conditions very different, the demand is much more indeterminate, much more elastic. Specification may be anything from a speculative interest, or a request for an equipment to reduce time, cost, or risk, to an order for a standardized package. It is, moreover, not merely a question in the former case of selling development work as a service; what is very often required is a combination of technical study of a potential need, and interpretation between need and available technique. In one firm, it was argued by development engineers that for any equipment which is a technical novelty in the market, as well as for equipment in the development stage, the development engineer himself has to educate the customer, to 'persuade him that he is interested'.

What is the role of sales in this kind of situation—apart from the kind of social obstetrics traditionally performed by salesmen, which may even be merely grotesque on occasions dominated by technical problems, and apart also from the formal conclusion of contractual agreements? As products approach some standard design form, or as applications become determinate enough for customers to be able to provide clear specifications, the salesman can assume more and more of the basic sales functions, but there is a large and important sector of sales activity in the electronics industry which is concerned with the conception and embryonic development of new ideas for designs and is as distinct from normal selling operations as research and development is from standardized design work. It is further complicated by the fact that normal selling operations should be continually invading this forward sector.

The move out from a market dominated by defence ministries has, inevitably, produced a piecemeal, somewhat confused diversity of conceptions of the sales task and of the structure of the sales organization. The commonest immediate reaction was to strengthen the sales force and raise its standing in the firm. Sales managers were given places on the board of directors, or new commercial directors were imported. Development engineers with the right aptitudes were encouraged to become technical salesmen. The managing director might himself act as chief salesman. In two cases, large firms reorganized their whole structure to give due prominence to their new commercial purposes.

Among the varied responses made by the established electronics firms to the new market situation, five kinds of solution may be distinguished:

(*i*) Common to almost every firm, and often in alliance with other solutions, was the transfer of technical staff to the sales force. This clearly promotes the purely linguistic ability of sales to handle new designs, but equally clearly yields diminishing returns as design advances, and can lead to an aggravation of the difficulties if salesmen find it difficult to acknowledge their gradual loss of technical mastery.

(*ii*) The course followed by the larger concerns, which are dominated by the need to sell a very large aggregate production load, has been to acknowledge this dominant need by creating powerful sales organizations; the increasing specialization of knowledge both of technical possibilities and of user requirements demanded by continuous design progress is allowed for by splitting the sales organization into special product groups. Each of these is served by a development design laboratory. The leading edge of the product division's activities is selling; the divisional manager is nearly always a salesman. Major difficulties have arisen because, simply, the new structure of product divisions is not something brought into being where nothing existed before, but signifies the reconstruction of the old order in which the development-design laboratories were usually the most powerful or the highest ranking group in the company, and always took precedence over sales. This supremacy was now challenged, if not overthrown. Reorganization according to the new arrangement provided a convenient arena for conflict between salesmen and engineers for the control of resources, policy, and patronage within the product divisions, and throughout the firm. Many such issues were in active debate at the time the study of these firms took place. As we shall explain in later chapters, these conflicts played a significant part in shaping the organizational structure. To take a simple example, in one firm three distinct groups were occupied with computer development, two at different removes from control by the manager of the computer product division; two of these groups were designing and making for the same customer.

In addition, product divisions themselves competed with each other for development and production resources. This reinforced the natural rivalry between them for business, a rivalry which, it was said, inclined their principals to measure achievement in sales volume rather than in profits. 'Product divisions', said the accountant in one large company organized on these lines, 'will always tend to sell in the cheapest market, to build up sales'. For in this way they could reinforce their claims for more production resources and services—and capital expenditure—to be allocated to them.

(*iii*) In some firms, the problems and importance of the shift of balance towards the commercial market prompted the managing director—or the chief executive—to take on the chief sales job in addition to his own. Doubling roles in this way and at this level always produces delicate situations—for others, if not for the managing director himself—prevents him from exercising over sales the same directive control as he exerts over other functions, and suppresses the critical debate between sales, design, production, and financial control out of which policy decisions emerge.

(*iv*) A fourth policy was to acknowledge fully the commercial role of the development design engineer and to commission him to explore the market for new user needs which might exist for application of the new ideas formulated in his particular technical field. Such exploration may involve protracted periods of operational study of user needs, and full responsibility for making, selling, and installing the first 'prototype' models of the eventual design. Just as the production side would take over from the laboratory and model shop after a design had advanced to this stage, so did the sales organization take over the large-scale exploitation of the potential market.

In one case, it so happened that the initial product range of laboratory instruments impelled the firm in its early days in the direction of very close liaison between designers and users. This relationship has since been preserved as of fundamental importance, and indeed the distinctive feature of the firm was the successful harnessing of 'R & D' effort to commercial ends. This has been done by directly involving laboratory engineers in the exploitation of user needs, in market survey and in market development for products such as business machines and industrial controls. The consensus of opinion on the need to think of development projects in terms of profitability was very striking; it was, for example, quite outside experience elsewhere to find a laboratory project leader beginning an answer to the stock interview question 'What do you think makes a good industrial scientist?' with 'Salesmanship, in the first place. . . .'

One other firm had consciously translated the relationship between development engineer and user, which they had learned in the context of government work, into terms appropriate to the commercial market. Again, the build-up started from a thorough exploration of the potential market—this time for radar navigational aids. This led to a specification of the equipment and of a price limit, to a design appropriate to the specification, and eventually to a production programme which would enable the price to be met. But the crucial principle to which the company adhered was that of regarding its total task as that of discovering user needs and interpreting them into products with the techniques and resources available to it.

Both these firms have, moreover, made it one of the cardinal features of their policy, since their foundation in the post-war years, to avoid becoming dependent on government work. Both have created very large international sales networks, and there is a conscious driving of the whole organization from the market end. Both have applied themselves to the systematic study of user needs, searching for gaps in the market which available techniques might fill. In one of them this exploration is the responsibility of the development side. What is more, manufacture and sales of early prototype models are their responsibility; and the return from such sales is credited in their budget to the development laboratories.

(v) Lastly, there is the possibility of creating for the sales organization itself a commercial policy-making function, in a technical sense. This was the least favoured policy, since what is involved is some transfer of the central functions of top management (i.e. of the managing director and board). What is also involved is a translation of the mysteries of 'market flair', 'business acumen', 'intuitive awareness of the way things are going', 'ability to spot winners', and all the other quoted characteristics of the successful industrialist into terms of forecast studies and analyses of technical, industrial, market, and consumption trends. The two points of central importance, however, are that this function has to be discharged in a comprehensive and continuous fashion over the whole market, present and future, actual and potential, of the firm, and that the discharge of such a function permits sales to occupy a technical, information-providing role which might furnish a basis of co-operation instead of competition with development and design.

In neither of the two cases, however, in which this distinctive function had become clearly visible in a working group devoted to it, did it form part of the sales organization. Significantly, both groups were regarded as the organizational pets of the managing directors, at whose instance they had been created. In one firm, even more significantly, the managing director's purpose in forming the group had been to ease his own burden, and to shift the centre of gravity of job market assessment and commercial policy making away from himself. Hitherto, the very rapid expansion of the company had been guided by a small group of ex-R.A.F. engineers whose wartime experience had been directly concerned with the technical, two-way, interpretation of operational requirements and technical possibilities in ground radar. To this fundamental skill they added ability to 'talk the languages' of naval, air force, and marine users of radar equipment. 'In the last ten years', said the managing director, 'we have travelled around, built up an intimate feel of their problems in a large number of countries, and are able to translate this into technical possibilities, even to the extent of seeing what will be needed in the future.' 'We', in this context, meant the managing director himself, leading a team of chief development engineers, design engineers, and technical sales managers. With the multiplication of products and markets, and with the growth of the company, personal guidance by the head of the concern and the recurrent absences of large fractions of the top management had become no longer feasible. So the company had begun to build up its 'operational research' group, charged with identifying technical trends and market trends, and with accumulating information about user needs and potential user needs. In order to provide the necessary occasion for contact with customers, the section was responsible for all enquiries and complaints about equipment already in service: the section closed the circuit between the consumer and the firm. In the other firm, a similar group had more of a business economics bent, and had no executive function; it provided a kind of internal business research and consulting service. Assessment of the market for computers, for example, and of the trend of computer development among bigger concerns in the industry, led to the firm's deciding to explore the needs for equipment to handle the input and output of computing data.

Caught between their relative unfitness for selling in the open commercial market and the shrinkage of the government market, most of the eight English electronics concerns were guided by a traditional conception of the salesman as the 'outside man', specializing in handling customers; as their view of the importance of this specialist job grew, the more inclined they became to alter the structure of their companies so as to put the specialist sales function in control of other specialist functions.

In the two firms we have described which were manifestly dealing with the new situation most successfully, the sales function of the development and design engineer, which was part and parcel of obtaining and working on Government contracts, was explicitly recognized and adapted to the new kind of customers, who had a wider, international, choice, and less concern in leading and nursing the industry. Under these new circumstances, the sales task in general was seen in such firms as relevant to all sections of the firm, and each specific job of negotiating with customers and exploring the market for new needs, as something which could be discharged best by those who developed or designed equipment, or by specialist salesmen, according to the degree of technical novelty involved. As one managing director put it, there tended to be a kind of oscillation between design and sales. 'As work piles up in the labs., you find engineers tending to demand that sales people and the factory must solve their own problems, while they tackled their jobs on hand; at other times, you get them complaining that sales were encroaching too far on their area of customer-contact; they were mishandling negotiations, they were letting the labs. in for impossible commitments.' This, he thought, was the way in which the strategy of the firm's sales became fitted to the amount of technical novelty involved in the firm's products at any one time.

Domestic Radio and Television Receivers

The products of the eight English firms were sold to three kinds of customer: British defence ministries; industrial or business undertakings and foreign governments; and the audience for radio and television entertainment. Each market has its special requirements and its own sets of conditions. Each, moreover, strongly influences the internal organization of the concerns which supply it. Nowhere was this more evident than in those concerns for which television and radio sets were the main product.

While the domestic radio market has grown, and manufacturers with it, competition has nevertheless remained keen; the continued decline in the number of producers is a fair indication of the market conditions. It has, in fact, always been one of the toughest mass markets, although the reasons for this have changed since before the war.

Conditions may be expected to become even more stringent with the inevitable flattening out of the demand curve as the home market for television sets becomes saturated. There was little concern about the future among managements, however, the chief expressed reason being the confident hope that colour television and multi-channel sets would keep demand up to at least its present amount. There was also some hope that Commonwealth markets, in which one or two firms had a footing, would expand. One firm was even convinced that its goodwill with dealers and reputation in the trade provided a material insurance against any severe turndown.

The typical reaction, however, was:

'When the curve flattens off, there will probably be colour, there will probably be alternative programmes which will go into different frequency bands and stimulate development in that respect. That sort of thing will take you—what? another ten years, anyway. That is what people have got at the back of their minds, although you wonder occasionally.'

More important than the explicitly stated reasons seemed to be the fact that the middle-distance future was obscured by extreme pre-occupation with the short-term future. Firms in the radio and television business are geared to annual cycles of new designs—production—sales, each successive activity rising to a peak in the early spring, the summer, and the autumn.

Three concerns in the study made domestic radio and television sets. For two of them, this was by far the biggest business interest. The third concern was a plant equally dependent on radio manufacture, but a subsidiary interest of the very large concern of which it formed part; a section of the head office sales organization, geographically remote and organizationally quite separate, sold the sets they made. The first two were, on any showing, far more successful than the third. And other clear distinctions were apparent between the first two and the third.

In the first two, every section of the management was continuously and actively aware of the dependence of their concerns on competitive selling. In both cases, the managing director kept in close personal touch with salesmen and even with retail dealers. Both men were credited with superlative commercial flair and market judgement; these qualities were regarded as the keystone of the firm's existence, out-weighing the information which the sales force and market research might produce. Nevertheless, all the information, skills, technical knowledge, and judgement of the firm were contributed to the focal task of determining sales policy, and especially of deciding on the outward appearance, performance standards, and price of sets. The centre-piece of the management system was said, by the chief managers in both firms, to be the weekly meeting of directors and chief executives to discuss current sales, the current production programme, sales strategy and forecasts, and new designs.

While the products of the radio industry are technically the most complex article on sale to the public, they are sold almost entirely on appearance and price. This is so fundamental a part of the industry's life, for these two firms, that the top management weekly meeting devoted half its time, throughout the year, to the styling, price, and cost of new models. Elaborate mock-ups of cabinets would be prepared for display at the meeting, amended versions would be made with equal care. Opinion canvassing would go on outside the meeting among salesmen and other juniors. Designers and production engineers would be called in to discuss technical points. Indeed, the formal, weekly 'top management' meeting went on in the context of constant consultation between executives at all levels.

Such meetings had an expressive, almost ritual, function as well as their manifest operational task. They signified the dependence of all sections of the enterprise on the successful accomplishment of the commercial task of the firm: of designing and making sets that people would want to buy. Even the works manager in one firm, asked what he thought was 'the main requirement' of his job, replied 'awareness of the market'. Design engineers were still more specific about the rooting of their job in designing for the market. The meeting signified interdependence, too. There was something of a ceremonial reaffirmation of solidarity and common purpose in the Monday morning assembly of heads of each major sector of the concern. It reaffirmed also the need for agreement and co-operation, if the firm was to survive. Functionaries imbedded well inside the working system had each week to emerge from their exclusive preoccupation with the tasks and problems and forces acting on them from other individual groups inside the system into a consensus about the reason and basis for the firm's existence. So, a chief planning engineer could say that his people regarded it as their job to 'overcome any amount of trouble if it was for the good of the set'.

Operationally, the importance attached to the meeting was implicit in the crucial nature of the decisions about design, prices, and programmes made by the meeting. Formal authority for the approval of models and for the issue of the works orders which set the machinery of production in motion rested, in one firm, with the meeting as a whole.

In the third concern, sales and design-production were independently organized. There was a monthly sales meeting. The chief television designer did not attend it. Until recently, model style had been decided by the sales manager choosing among coloured drawings. There was no arrangement for contact between designers and dealers; trade reactions were interpreted by sales and passed on to the management of the concern, and it was said that remarks of girls on the assembly lines could—and did—provide a quicker and more accurate appreciation of the saleability of a model than the design works and management ever received from sales. Other evidence all contributed to the impression that the organizational and geographical separation of sales from the rest of the system was the most important of many circumstances which combined to hamper the commercial effectiveness of this particular concern. This impression was confirmed in later discussion by the principals of the firm.

SELLING AS A TASK OF THE WHOLE CONCERN

This review of firms and the three kinds of market has arranged both according to a rough gradient, from the treatment of the market as a sink, into which a firm tries to pour applications of techniques known to the firm, to treatment of the market as a continuing source of needs, actual and potential, which the firm tries to satisfy. Whether a firm treated the market as a sink or as a source appears to have been the most important determinant of its success. The predominance of one attitude or the other is tacitly but unmistakably revealed in the way the firm has organized its dealings with the market for its products. The overwhelming importance of the Government market in the formative years of the post-war electronics industry reared many firms in ignorance of the realities of life in ordinary competitive markets. The firms with the fastest rate of growth during the 1950s are those which deliberately avoided too large a commitment to government work.

If the firm regards itself as a system which is being 'driven' by its market, then the sales function, in its very broadest sense, is something discharged by the whole firm. This is patently obvious in the case of the efficient firms in the radio industry, which produce standard products for a mass market. We have tried to demonstrate that, in such firms, this orientation of the whole system and of every job to the sales function is not incompatible with the fact that the task of interpreting the market to the firm and selling to the thousands of retail dealers who are the firm's customers is the affair of a sales department. At the other extreme, the sale of complex and expensive telecommunications or radar defence systems to a foreign government might be handled by systems-planning engineers because neither a salesman nor an engineer concerned with designing individual bits of equipment could formulate a comprehensive answer to the problem posed by a customer. In between is a broad field comprising the sale of equipments such as scientific instruments, computers, and radar apparatus, designed for specialized purposes and made in small numbers; here the sales function might be discharged in the initial phases by design engineers, with salesmen taking over as demand and design stabilized. In the firms with the most successful sales record, the sales function flowed most easily between development and design engineers and the sales organization. So the most prosperous and fastest-growing radio manufacturer sold radio and television sets through a sales organization (but held its weekly meeting of directors and heads of departments about designs and sales) and had the engineers responsible for the development and production of semi-conductors in a new subsidiary company themselves carry out the surveys of the market for their products.

NEWCOMERS TO THE INDUSTRY

We were very early on made aware of the important and pervasive influence on the organization and conduct of management which derives from the way a firm sets about achieving its commercial purposes in a market. At the outset of the Scottish study, we tried to ascertain the incentives which had encouraged the firms to come into the Scottish Council's electronics scheme. These firms were all small or medium-sized, employing from about 150 to about 1,500 people. Apart from the trying years of 're-conversion' in the immediate post-war period, all were reasonably successful according to the criteria usually applied to business concerns. Being in Scotland, they were some distance from the main centres of industrial innovation in most fields, and especially in electronics, which is heavily concentrated in and around London. They were, geographically, country—even back-woods—members of the 'electronics club'. This undoubtedly hampered the firms which did not see this as something they had to overcome by positive effort. It made directors extremely cautious about commit-ments, rendered them over-dependent on the goodwill of defence ministries, and made it harder for them to develop effective relation-ships with the markets already receptive to electronic devices.

In every case, the primary factor in the firm's decision to explore new technical ground in order to derive new products was the shrink-age or closure of the market for its existing products. In some cases, the 'red light' had been visible even before the war. One company had manufactured railway rolling-stock for three generations. During the thirties it had got into low water and there was a real danger of its shutting down altogether. A large Scottish industrial group bought the controlling interest, and kept it in being. But the manufacture of railway carriages and wagons was clearly becoming less and less profitable and, more important, less and less secure. The wartime expansion of demand saved its life, but was never regarded as affording the firm more than breathing space. And after the war, the trend to standardization ren-dered the firm still less able to compete with larger concerns which had invested large sums in new plant capable of handling large lots of one-design all-metal vehicles.

Another firm had been started during the war, at the instigation of the Government, and had operated exclusively on Government pro-duction contracts for specialized components. At the end of the war this work virtually stopped, and the management cast about for a basis on which to continue in peace-time. The decision made was to enter the commercial radio receiver market. This was later seen to be an entirely mistaken decision, since the firm was far too small for an essentially mass market, and because metals and other materials were scarce; what there was went to the bigger concerns who had been in business before the war. There were periods of weeks when work was at a standstill while assemblies awaited minor parts. This episode lost almost all the money made through the wartime production con-tracts.

A third firm in the electronics scheme had been a medium-sized engineering concern with a long tradition and wide reputation in precision mechanical engineering. During the war it had expanded production of its former products, but afterwards faced a situation, both in Government and commercial work, in which its high-precision work was becoming outmoded by electronic control and radar devices. It was only after some years, when the scale of the firm's activities and labour force dropped very considerably, and when a good deal of rather unrewarding jobbing work was taken on in an effort to stave off further decline, that it finally took the plunge and engaged a group of electronics development engineers.

Another, also with a long tradition of precision instrument manufacture, followed a roughly similar course of investing in development —although on a much smaller scale—in anticipation of innovations which would help it to recapture or improve on its former position in the same market. Its management had 'received an object lesson in just how quickly you can drop out of the market when you decide not to spend money on keeping in it'. This they had found at the end of the war to be true of one of their two principal products; they had decided that the expense of buying new patents and getting in the machinery and equipment necessary for improved and cheaper production would not be worth the larger slice of the market which they might hope to get.

The largest of the plants involved in the scheme was a long-established electrical engineering firm producing three or four fairly heavy and complex pieces of equipment for an international market. The firm was expanding, but had become increasingly aware of the growth of research and development resources in other large firms in the industry. This disquiet came to a head when one of its products, derived from rights in foreign patents, became obsolescent through the development of techniques outside the range of the firm's competence. Partly to make good the loss of production and to keep staff, but mainly to acquire some of the technical information and skills which the firm lacked and to make a start with the research and development work which they were now assured it should include as part of its ongoing activities, the management negotiated a development contract with the Scottish Council, Ferranti Ltd., and the Government, and began recruiting development engineers.

The other firms included in the Scottish study followed much the same pattern. Naturally, many other firms have found themselves in similar circumstances. Naturally, there were other courses open to them than entering electronics development work. These firms, however, did enter the Scottish Council's scheme. They did so partly because of the special knowledge or enthusiasm of a managing director or an influential member of the board, partly because the Scottish Council's scheme reduced the initial risk, saved time, and offered some guidance in a new territory. There was also the attraction of the prestige industrial research and development was now beginning to carry.

At the outset of the new phase of activity, some firms ventured into unfamiliar markets either because they thought 'there was a market somewhere' for innovations which they might be able to develop or copy or because they thought there was room for newcomers. The Ferranti case (pp. 55–6) is just such an instance. They fared badly and withdrew after losing money, without having acquired sufficient familiarity with the market to exploit it further.

Despite the dominance of market considerations in their decision to enter electronics, none of these firms matched its technical expansion with anything like comparable expansion of its sales activities. In the two cases where some success was achieved, the head of the development group himself was allowed, and later encouraged, to find customers and explore requirements which his group might meet. Apart from these, no attempt seems to have been made to survey the possible demand for articles which the firm proceeded to develop and make. Instruments were produced which were found to be obsolete or commonplace in the specialized markets to which they were offered. One of the firms developed early in its new career of innovation a commercially marketable product which received some recognition as a particularly successful design at a trade exhibition; yet nearly all enquiries from potential customers were answered by correspondence for a period of many months, and years went by before a technical sales programme was developed.

All firms to begin with, and most at later stages, curtailed the activities of their development groups by keeping them at work on specific items of known user demand. Extreme cases were those firms which virtually refused to consider using their development teams for anything other than government defence contracts, few of which even led to production contracts. Thus, in addition to limitation of the market possibilities, there was limitation of technical possibilities.

Four firms negotiated a defence development contract before engaging a development team. These firms were in general those which restricted most narrowly the technical activities of their development groups so far as commercial user needs were concerned.

Market difficulties were not the only troubles of these firms, where in many cases development work was dropped after an unhappy two or three years. But they were fundamental, and they were the most important aspect of a tradition of management which proved itself utterly inappropriate to the needs of industrial undertakings faced with changing conditions in their technical foundations and commercial tasks.

PART TWO

Organization and Change

CHAPTER 5

Management Structures and Systems

ORGANIZATION AS AN INTERPRETIVE SYSTEM

The twenty concerns which were the subject of these studies were not all separately constituted business companies. This is why we have used 'concern' as a generic term. Some, although employing many thousands of workpeople, were only small parts of their parent organizations. Others comprised several separately incorporated companies, all, however, subordinate to a 'main board' or parent company; they were regarded as single businesses carried on under what was virtually the same top management group. Our unit is an organized working community, concerned with designing, making, and selling products which incorporate electronic components. In those few cases in which a part of a 'head office' sales organization sold the articles made by the concern, the salesmen have been treated as part of the concern; similarly, accounting control and other commercial and financial functions carried out by a 'head office' organization through, or on information received from, a subordinate office in the concern, have been regarded as within the concern. The titles of the heads of the concerns were variously chairman, managing director, or general manager.

Each concern comprises many hundreds or thousands of people who, when they are at work, are simultaneously engaged in a great number of different activities. All these activities are connected, and indeed interdependent. But they are not all directly connected at any one moment. The instructions being written out on a planning schedule do not refer to the operations now being performed by any machine or operative; the terminology used, and the subjects discussed, by the group of laboratory engineers talking over the power supply needed for an electrical circuit are different in important respects from the language and matter of the overseas sales conference proceeding in the managing director's office. The sheets of steel now being bent, cut, and punched in the metal shop may be quite different in size and shape, before and after, from the chassis and containers now standing, encrusted with electrical components, on the benches in the assembly shop.

The connexion between all these activities is nevertheless clear and direct. They have a common purpose, and they follow upon each other. Different as are the physical actions, manual skill or dexterity, intelligence, scientific knowledge, business experience, social finesse, and so forth which are employed in these activities, they have also a formal resemblance: each is performed in response to information received; each involves altering, rearranging, or recomposing information or things; each ends with the transmission of the altered information or thing to somebody else. The 'information received' may be anything from the visible presence of bits of material at the side of an operative's bench, or within reach on a moving belt, to the contents of a day's conference with Ministry officials about guidance techniques or a managing director's remark to the effect that 'we ought to start thinking seriously about colour television'. 'Acting on this information' means the application to it of technical or manual skill, or knowledge or understanding, in conformity with the expectations of the other members of the organization. The process, in so far as it is isolable as single actions, is complete for one individual only when information is transmitted to all the destinations in the concern, or outside it, where it is needed for effective sequential action.

We may in fact regard a manufacturing concern most simply as a device for translating orders for goods, contracts, or user demands in general, into articles or services. The whole process of translation is broken down into a series of steps, each of which is in itself a translating operation. Put at its crudest, the factory receives at one point orders, in the form of letters, telephone messages, schedules, instructions: a communication in language of some sort. From this point, the order goes through a series of mutations; in an engineering plant, for example, it is translated into a design and a specification; the design is translated into a number of orders for parts and materials; at that stage the subsidiary orders are translated into actual bits of material by the stores; other bits of material are translated in the machine shop from crude blocks of material into components; in the assembly room stacks of separate parts are translated into a number of finished articles; and so on, until the final stages of packing and dispatch.

ORGANIZATION WITHIN A STABLE PROGRAMME

It is not, of course, just as simple as that. Complications enter in, but we can best show how they do, and their consequences, by taking an actual instance. For this, it is convenient at this point to go outside engineering and describe the working of a concern in a process industry, the subject of the first preliminary study (see pp. 1–2). The plant produced viscose rayon filament yarn, and employed about 900 persons. It was owned by a company with offices in London and one smaller factory in Lancashire. Sales and costing were operated by the London office, but all other departments of the firm were represented in the factory.

The production process proper starts with the intake of raw material into the stores. The principal materials are cellulose, carbon bisulphide, and sulphuric acid. The first stage is the arrangement of the stored cellulose sheets into batches of uniform weight. One man and a girl do this, with no mechanical aids apart from a weighing machine and a hand trolley. Although this stage can be referred to as a single interpretative step in the series, it obviously can be broken down into a large number of detailed movements; the essential point for us is that each separate movement is almost literally a 'translation', even to the transport of a batch from the platform of the weighing machine to the place by the end door where its arrival denotes its readiness for the next stage. This involves handcarting to a vat of caustic solution in which the batch is steeped for a period; at the end of this time, the cellulose is pressed between plates by a hydraulic ram, then carted and fed in to mechanical kneaders. The shredded cellulose is then stored for two or three days in mercerizing rooms in order to ensure a completely homogeneous moisture content and to prepare for the next chemical process, which is mixing with carbon bisulphide in mechanical churns. After this, the product, cellulose xanthate, is dropped through chutes into mixers, in which the compound is dissolved in a solution of caustic soda and mixed by beating and forcing through blend jackets into a homogeneous fluid. The viscose is stored under vacuum to remove bubbles, filtered, and then stored again in charge tanks.

All the stages up to this are preparatory. The next stage is central to the whole process. It consists in pumping the viscose through a multiple jet into a bath of sulphuric acid, from which it emerges as solid fibres. These are drawn up over a wheel, twisted to form a thread, and passed down into a spinning pot, in which the thread is built up into cylindrical cakes. The finishing processes consist of washing the cake with neutralizing solutions and water, drying it in heated lockers, inspecting, and so on to weighing and dispatch.

So far we have included in the system no more than certain muscular and mechanical efforts and chemical processes. At the same technical level there are a number of actions which are not themselves part of the processes of change, as we have so far noted them, but are essential to their proper performance. The spinning machine, for example, is automatic, but a number of workers are present renewing broken threads, removing completed cakes, and replacing empty pots, an operation which involves stopping and starting the machine. In the mercerizing room, heat and humidity are controlled by thermostatic and other apparatus, which also keeps a record. A group of workers are employed to inspect these and other temperature and humidity records and to report fluctuations outside a certain latitude. In addition, there are certain routine sequences of actions proceeding at regular intervals. The baths of sulphuric acid in the spinning room are sampled every hour. The sample is taken to the laboratory and analysed. Any fluctuation beyond certain limits is noted, and instructions delivered to the foreman in charge of the 'acid house' to increase or decrease the input of sulphuric acid.

In these instances, all action is designed to produce as much rayon as cheaply as possible, within the limits of the resources of the company and the requirements of the market. These are clearly defined in a programme. The normative character of everybody's work is quite explicit in the factory. The job of the 'efficiency man' in the spinning room is to approximate the process of transforming raw viscose into yarn as closely as possible to complete effectiveness, in which no machine time and no material are lost. The other control operations mentioned have the purpose of keeping the process changes and the conditions affecting those changes within a range of limits, and close to a series of constants. There is, in fact, a collection of permitted tolerances ('limits and constants') set down for all stages, which are bound together in a book. This was called the 'Factory Bible' and was in the hands of every head of department. Most of the skilled work in the factory, and a good deal of the work of foremen and heads of departments, is the control of processes so that they act according to the norms laid down in the 'Factory Bible'. Systematic control of this order can be regarded as a series of circuits. Some are extremely short and entirely mechanical, as when a deviation from a specific temperature is corrected by a thermostat acting on the supply of hot water; some can be short and operated by a single worker, as when a clogged jet is replaced and the thread taken up and guided along its path into the spinning pot; some can be fairly complicated and involve a number of men and operations, as in the case of the routine laboratory tests.

Control procedures at the higher executive level are necessarily complex, and this is not the place to give a detailed survey of them. In this rayon factory the most elaborate of these procedures are those which relate the production process to sales requirements. At the outset of this description the receipt of orders was placed at the beginning of the whole production process as the logical precedent to it. So it is, but since the whole production process takes anything up to eight or nine days, contingencies are continually arising which make it necessary first to alter the production process, and consequently the amount or nature of the product, and second to restore production to the norm from which it departed. For example, the length of time viscose stays in storage before spinning affects its quality. Other conditions also affect quality, but are linked to the period of storage. Tests of quality are continually made; and if the storage period seems likely to become excessive, the usage of viscose is speeded up by thickening the denier of yarn, i.e., the weight per yard; this means that only the heavier deniers ordered by the sales department are produced for a time, and adjustments have to be made later either in the spinning room, in the earlier parts of the process, or to sales action itself. Most of the time of the Works Manager is spent in reconciling actual production with maximum potential and with required production—three variables each of which is recorded elaborately in logs, production forecasts, weekly programmes, and reconciliation charts.

In turn, the General Manager reconciles variations in the performance of the production departments with fluctuations in the contribution of the service and maintenance departments, chief chemist's section, personnel department, and 'reconciles' the total outcome of the works' activities with the expectations of the head office. These expectations are expressed as figures of programmed sales, programmed production, and costs. The whole concern is visible as a pyramid of knowledge about the circumstances of the concern. As one descends through the hierarchy, one finds more limited information, technical and local, about these circumstances, and also more limited control over the resources of the firm. One also finds each person's task more and more clearly defined by his superior, so that he is capable not only of knowing what to do in normal circumstances without consulting anyone else, but also knows just how far he may allow a situation to depart from the normal. Beyond a certain limit he has insufficient authority, insufficient information, and usually insufficient technical ability to be able to make decisions. He is informed, therefore, quite clearly when this limit occurs; beyond it, he has one course open—to report to his superior. Similarly, his part in the common purpose is defined and it is normally unnecessary for him to have to consider further how his task relates to the firm's commercial ends.

Although the specification and detachment of the individual member of the organization increases the lower down in the hierarchy he is, even the General Manager's task is carried out within the framework of a programme—indeed, of a very precisely defined programme, with uncertainties and expectations, so far as demand is concerned, ironed out for him beforehand.

The system of management within the factory was quite explicitly devised to keep production and production conditions stable. With this as the underlying principle, the system defined what information or instructions arrived at any one position in the hierarchy, what information or instructions might leave it, and their destination. Such definition was a matter of fixed, clear, and precise routine. Similarly, each working position in the hierarchy had its authority, information, and technical competence specified once for all. Moreover, since each position below the General Manager's in the hierarchy was specialized in all three features of authority, technique, and information, and nobody was empowered to act outside defined limits, all departures from stable conditions were swiftly reported upwards, and, so far as the works were concerned, the General Manager existed as the fountainhead of all information about commercial and other conditions affecting the affairs of the factory (as against technique). Such changes as did occur, therefore, were inaugurated at the top. There was, accordingly, a fairly stringent authoritarian character about the conduct of superiors to their subordinates, an authoritarianism which was accepted as reasonable, and did not in the least interfere with sociable friendliness on an equal footing on the many occasions on which members of the staff met each other outside work in the small town in which the works was located. It was perfectly possible, that is, for members of the firm to accept instruction and command as appropriate to work relationships, but to isolate these relationships from what went on outside.

What we have been considering, in this example, is the structure of management in a concern for which technical and market conditions approximated very closely to stability. At all levels, decision-making occurred within the framework of familiar expectations and beliefs, many of which could be formulated numerically as a programme. Fluctuations in demand did occur, but these were treated as deviations from normality, and part of the task of management was to constrain the sales office in London to avoid such deviations. Production pro-grammes, which in operational terms were planned for a week's run, were devised for monthly, three-monthly, and six-monthly periods so as to make it easier to run each weekly programme without alteration.

ORGANIZATION FOR CHANGE

We may next consider the way in which the management of another concern operated, this time in the electrical engineering industry. It too had a hierarchy of management, most of whom performed speci-alist tasks. But although, in this instance as in the first, preliminary interviews with the head of the concern began with a chart showing the hierarchy, the sketch drawn for the interviewer petered out in un-resolved dilemmas. There were two separate plants in the company, near each other, each concerned with the final production of one of two types of switchgear. Design and production was on fairly standard lines—this was, in fact, a principal endeavour of the management, and a main factor in the commercial success of the firm—but every contract required some special units, and there was a constant flow of design improvements. There were frequent episodes in which changes in the scale or kind of operations being performed by one manager reper-cussed through the structure. The Production Controller in Factory A recounted one such episode:

Each factory performed certain processes for both—all turning, drilling, etc., was, for example, done in one, and all sheet-metal work in the other; the same Works Manager supervised production in both, and buying was also centralized in one section. Each factory, however, had its own production controller. Some months before our first visit, the production controller in Factory B had decided that the stock levels of a number of parts for his product were too low and had raised the minimum stock figures on the stores cards. Orders for making these parts in numbers sufficient to meet these higher levels had been passed in routine fashion to the machine shop in Factory A, which had consequently found its monthly output programme abnormally high. Orders for metal stocks also rose, later on. The pressure on his machine-shop and the rise in material costs had mystified Production Controller A, since they were not apparently related to a comparable rise in the forecast of final units to be produced in either factory. Eventually, of course, Production Controller A had run the thing to earth, but there had been a period which the progress of work through his factory had been out of gear, when he had been unable to explain whence the increase in machine-shop time had come, and when he had been made to feel inefficient.

The recital of the salient points of this episode in this condensed fashion perhaps renders the difficulty too trivial and the solution too obvious. The common-sense response, only too clearly, is that if Production Controller B had only told Production Controller A what he was doing with the stores cards no trouble would ever have arisen. But difficulties of this kind form the everyday experience of industrial management. The recognition of this is contained in the widespread currency that has been given to obscure propositions to the effect that 'good communications are the essence of good management'. The point we are trying to make here is that with young, efficient, well-qualified, and ambitious managers in both factories, it yet did not occur to them to see that such information was passed; indeed, there are reasons against their so doing. Each person concerned was acting well within the boundaries of the rights and obligations which any formal definition of his function could have laid down. Stores records of odd items were fairly frequently altered by each production controller without any difficulty being caused; the machine-shop was a little too remote from Production Controller B, who had altered the cards, for him to foresee any noticeable dislocation arising from his action, an action which was, in any case, the consequence of a continuous rise in production figures which 'everybody knew about' and which must have been to him an obvious source of rises in demand for parts and stock. The trouble arose *because* the persons concerned were acting within definitions of their roles which were assumed, or were implicit in an undisturbed routine, or had even at some time been explicitly stated.

The point of interest, however, lies in the fact that, in this company, procedures were devised to meet contingencies precisely like this. In fact, the whole hierarchic structure of management seemed to observe different principles from those of the rayon mill. It was claimed that nobody on the staff had a title except the managing director; at least, nobody had a definite function to which he could keep; this was first said to one of us, somewhat ruefully, by a senior member of the staff in the presence of other members and the managing director, who all agreed. The firm's own practice was not discussed, but was defended by implication in derisive accounts of other firms which made great play with title throughout the senior and middle management ranks. It was also an explicit rule of the firm that any member of the top management group could be consulted or asked for a decision by any junior. There was, therefore, a deliberate avoidance of clearly defined functions and of lines of responsibility at the top level of management, which, although to a lesser extent, carried on down through the rest of the management structure.

The firm had, however, developed organizational procedure of an equally explicit kind in order to co-ordinate the functions discharged by individuals and fit them as a whole to the requirements put on the firm. A management committee met every week. Normally this consisted of the two principal executives in the two main factories of the concern, the works manager, the production controller, and occasionally one or other of the specialist executives who might be particularly concerned. These meetings were additional to frequent inter-action between the senior managers, and it was said on more than one occasion that if, because of the absence of more than one of them from the works, a week went by without a meeting, they all felt the lack of it fairly strongly, and were relieved when the next meeting came round. At these meetings they 'found out what was happening and decided what to do about it'. Although certain decisions, like the purchase of machines, were for the committee as a whole to make, the function of the meetings was to enable each member of the senior management group to arrive at decisions respecting his own part of the firm's task in the light of information derived from the others and in the light of the implications of his decisions for the rest of the firm's activities.

In other words, the firm's expectations were constantly subject to alteration, and the framework of decision, consequently, was continually being re-set. Top management had, therefore, to interact continually in order to ensure a coherent and comprehensive framework for their own individual divisions and consistency between their several decisions. In this respect, the possession by all members of the committee alike of a common background training and language, in engineering qualifications, was held to be of the first importance.

But the dimensions of the Management Committee were not permanently set. It was regarded as a cadre which was capable of expansion at need into a large meeting consisting of almost the whole personnel of the management staff and representatives of the different shops. This happened when, as we were told, 'the firm seemed to be getting into the doldrums a bit', when overtime seemed to be called for fairly often, when production targets had not been approached closely enough, whenever, in fact, there appeared symptoms of dislocation between the programme and actual production, between earlier and later process departments, between one level and another of the personnel of the firm, and so forth. At such meetings they would usually find that after the preliminaries a good number of complaints and criticisms would start getting voiced. The first large meeting would be taken up with getting these ventilated, and by the time the second meeting occurred many of the difficulties behind the complaints would have been settled in the interval by people directly involved in them. The residue of difficulties would be discussed and possible solutions put forward. After a few meetings the volume of new problems which lent themselves to discussion by a large group of members of the firm would have dwindled, and the management committee would be shrunk to its ordinary size.

Meetings held 'to iron out difficulties' and 'improve communications' are fairly commonplace. The procedure described is nothing very novel or startling, although perhaps the retention of the title of Management Committee for a meeting open to so many members of the firm is unusual. What we wish to illustrate here, however, is the departure from the operating principles of a hierarchy of authority, information, and technical expertise which novelties, or fairly sudden alteration in the expectations and programme of the firm, may instigate. Description of two other kinds of meeting in the same firm may help to make this point clearer.

(a) Every week the Works Manager held a meeting with the foremen under him. They were affairs of a few minutes only, and consisted almost entirely of instructions about the week's production programme. Questions were asked and answered, and there might be discussion and even decisions by agreement about some matters such as synchronizing jobs affecting two departments, but the organizational conditions of these foremen's meetings were essentially those now familiar as the 'briefing sessions' of service units. The situations with which this group was concerned were clear, familiar, and stable; they were the routines of production and related to normal production factors. Anything abnormal, so far as the forecast week's production went, had been dealt with already by other people and had been translated into the normal and routine.

Nevertheless, the limits of the situations to which this kind of procedure applied were fairly strictly laid down. Apart from this programmed routine, the express policy of the company was for each foreman to regard the work of his shop as a contribution to the whole task of the firm. Thus work was wherever possible distributed between the two factories, and between shops, in a non-routine fashion, so as to avoid the growth of any feeling that certain parts only of the manufacture of products was 'our responsibility' and everything else 'somebody else's worry'. The interleaving of stages in the manufacture of each main product in the programme of work of each of the two factories and the retention of one Works Manager for both, even though they were a few miles apart, were meant to serve the same end.

(b) Soon after the firm had decided to start a small research department a committee was set up to discuss policy and general problems relating to design and development. This was convened by the chairman and managing director at intervals of five or six weeks, when it was attended by a research consultant. The Research and Development Committee had begun as a large, comprehensive affair, but it had shrunk to a normal membership of about a dozen. The reason for the reduction in size, we were told, was the substantial difference in levels of technical knowledge between higher and lower levels of the firm's hierarchy. It quickly became unprofitable to conduct the meetings in a language understandable by foremen, say, who were often not skilled craftsmen, and to take up time with the kind of problem which presented itself to them. Nevertheless, these development meetings were expanded to their former 'omnibus' capacity during a period when a major design change was developed for one of the two main products. The biggest single reduction of cost eventually effected by the new design proved to be in the direction of simplifying and cheapening the metal chassis and container, and the biggest contribution of ideas about this came from the 'shop floor'.

The Research and Development Committee closely resembles the Management Committee, and is clearly a different affair from the Foremen's Meeting. Indeed, the difference in the titles—'committee' and 'meeting' is significant. The Foreman's Meeting was no more than a convenient way for the Works Manager to see that routine instructions were disseminated to his subordinates quickly and uniformly and that they all understood what they had to do; each foreman knew the extent of the others' dependence on his shop during the week, and saw that the others knew what he was expecting from them. The hierarchic structure was working as a communication system, as it did continually in the rayon factory. In the Committees, however, the hierarchy of management and the clear specification of each person's authority, information, and technical standing was, to some extent, put aside for the duration of each meeting.

The constitution of a committee implies, in common usage, that each member enjoys an equal right to speak his mind on decisions and motions subject to the chairman's jurisdiction, and equal voting powers apart from the chairman's casting vote.* There are, of course, many kinds of committee for which this is only partly true; only rarely, for example, were votes taken formally in the two Committees in the switchgear firm. It is also to be expected that persons of lower rank in the hierarchy of management will exercise their right to speak in a more discreet fashion than their seniors, although the early experience of the Research and Development Committee indicates that the former were not over-cautious. Even with such qualifications, however, the presumption of equality must remain, since only in so far as the requirements of equal voice and equal voting rights are met is the committee an effective instrument. Otherwise, 'Why have a committee?' —a question ordinarily raised when these requirements are not met, or, being met, the opinions expressed or the motions carried are not be-

* This is the connotation throughout in Black's *Theory of Committees and Elections.*[31]

lieved to produce any of the effects they were aimed at* Thus, the Committees were devices for abrogating, for the length of each meeting, the distribution of authority, information (about the firm's resources, policy intentions, and circumstances), and technical competence pictured in the hierarchic structure of the management. The reason they were so abrogated is that the number of contingencies of new or unknown origin arising in the total situation of the firm in the field both of information and technique was at times too great for the management structure when it operated according to stable expectations and a programme. A new communication system was therefore called into being which used channels other than those laid down by the management structure. This was kept in being until the need for communication about non-routine local information and techniques slackened again to dimensions for which the system operated through the management structure was adequate.

The rayon concern had no need to resort to an alternative system, because of the rate of change in its expectations and programme was so slow as to be reduced to stability for the main part of the structure. In the switchgear company the rate of technical development and the repercussive adjustment in the commercial environment and in the internal situation of the firm (knowledge about which we have put in the category of 'local' information) was significantly faster. Hence the provision of an alternative system to what was normally inherent in the management structure. Moreover, there was a disposition towards departures from the routine, even for the conduct of ordinary affairs. This was manifest in the lack of titles, in the imprecise functions of managers, in the freedom of consultation encouraged between juniors and seniors irrespective of their departmental functions and authority, and in the importance attached to the Management Committee during the periods which most approximated to stability. It became more urgent on the occasion of committee meetings, but it was necessary at all times for each person to be aware of the common purpose of the concern, and the way in which his own part-functions connected with it, and with the contribution of other members of the firm. This was an essential condition for the perpetual canvass of ideas and information outside the limits ordinarily set for decision-making and the conduct of affairs in stable concerns.

* See, for example, the episode recounted on pp. 152–3.

ORGANIZATION FOR A CONSTANT OR PREDICTABLE RATE OF
NOVELTY

Even the switchgear company encountered no greater rate of change than would be familiar in the engineering industry at large. It was commercially successful and expanding fairly quickly, but on a basis of improvements in well-established design and of the low price obtained by extremely proficient production engineering and factory management.*

All electronics firms encountered much higher rates of technical and other change. At the same time, the resort to the 'alternative system' was more continuous, more pronounced, and more explicit. In radio and television manufacture, at the stable end of the industry's range, the operations of the firm are, as we noted earlier (pp. 67-8) dominated by management meetings. Such meetings are, however, only the formal elements of a comprehensive system of communication which involves the concern in the equivalent of a diffuse, intermittent, 'Management Committee' of the extended kind used for specific occasions in the switchgear company.

To take one of the successful radio firms, fortnightly meetings were held to consider 'forward designs', and these were attended by the Works Director, Technical Director, Chief Engineer, Sales Manager, and Publicity Director. Another fortnightly series of meetings, alternating with the first series, dealt with current production modifications and the progress of the new models in production. But this formal system of meetings existed within a context of constant consultation between the directorate, consultation which involved their subordinates. Design engineers and planning engineers were equally specific about the way in which consultation about design and styling stemmed outwards from the managing director, through the top management group, to include them.

Of the forward designs meeting, one director said 'We don't decide very much, but we help Mr. — (the Technical Director) to decide which kind of design to start. Sales have a lot to say about what they want, but nobody is dogmatic about it. I suppose the very fact that all this talk goes on with the Technical Director present helps him. You

---

* This judgement and those others which it has appeared necessary to make about the effectiveness of individual concerns are not ours alone. In every case, the statement reflects the opinion of other individuals acquainted with the firm, especially of competitors.

could say that all decisions are taken in the light of full knowledge of what other people think and of what they are doing. Similarly, they know about your decisions and what you are up to. This doesn't apply just to meetings.'

At lower levels in management the same free and frequent contact between individuals was maintained as prevailed among the directors. 'You go to the person who is most concerned with the problem in hand', whether foreman or director. 'If you want authority to get something done, you go to the top of the tree—no, not my tree— theirs.' Disputes with other departments could be referred to the directors, but these seldom arose, because 'When enough people are brought into it, problems settle themselves. Practically every question is settled logically (*sic*), not on somebody's authority. . . .'

Individual managers had developed their own insights about the effectiveness of the structure. 'When Bill in the shops comes across something that he doesn't like or doesn't understand, he can go straight to the design engineer or the draughtsman and whatever it is is sorted out at that level—that is, it's dealt with in operational terms between the people directly concerned. In some places, when this kind of problem arises, it has to be pushed up—i.e., to higher level executives, because the lower levels are out of touch with each other just as they are out of contact with the people at higher levels in other departments. So you can't get these diagonal contacts or these lateral contacts. You have to go through the higher levels, which means—and this is the most important point—that you have to involve people who are not conversant with the problem in operational terms. . . . There is hardly anybody ever involved in problems here who could be called non-operational.'

The consistent blurring of the definition given to individual positions in the management hierarchy becomes quite evident at this stage, especially as it begins to induce some insecurity. One manager, for example, said his 'chief problem' was that 'the limits of one's responsibilities and authority aren't defined'. There was no organization chart. In all interviews, managers found difficulty in saying who were 'at their own level' in the firm.

In the other major radio firm the 'Monday morning meeting' was an even more important element in the management system (see p. 67). It acted as a model for many other management meetings, many of which were also given executive powers. 'We are' it was said, 'always having meetings, because you don't bring about changes without meetings.' And change, it was unnecessary to say, was the normal condition of things for this firm.

ORGANIZATION AND INNOVATION

Beyond this point, in the electronics industry proper, one begins to meet concerns in which organization is thought of primarily in terms of the communication system; there is often a deliberate attempt to avoid specifying individual tasks, and to forbid any dependence on the management hierarchy as a structure of defined functions and authority. The head of one concern, at the beginning of the first interview, attacked the idea of the organization chart* as inapplicable in his concern and as a dangerous method of thinking about the working of industrial management. The first requirement of a management, according to him, was that it should make the fullest use of the capacities of its members; any individual's job should be as little defined as possible, so that it will 'shape itself' to his special abilities and initiative.

In this concern insistence on the least possible specification for managerial positions was much more in evidence than any devices for ensuring adequate interaction within the system. This did occur, but as a consequence of a set of conditions rather than of prescription by top management. Some of these conditions were physical; a single-storeyed building housed the entire concern, two thousand strong, from laboratories to canteen. Access to anyone was, therefore, physically simple and direct; it was easier to walk across to the laboratories' door, the office door, or the factory door and look about for the person one wanted, than even to telephone. Written communication inside the factory was actively discouraged. Most important of all, however, was the need of each individual manager for interaction with others, in order to get his own tasks and functions defined, in the absence of specification from above. When the position of product engineer was created, for example, the first incumbents said they had to 'find out' what they had to do, and what authority and resources they could command to do it.

In fact, this process of 'finding-out' about one's job proved to be

---

* 'Organization chart' is here used in reference to conceptions of management organization which identify the communication system with the hierarchic structure, with all that this implies (see pp. 104–6).

unending. Their roles were continually defined and redefined in connexion with specific tasks and as members of specific co-operative
groups. This happened through a perpetual sequence of encounters
with laboratory chiefs, with design engineers who had worked on the
equipment the product engineers were responsible for getting made,
with draughtsmen, with the works manager, with the foremen in
charge of the production shops they had to use, with rate-fixers, buyers,
and operatives. In every single case they, whose only commission was
'to see the job through', had to determine their part and that of the
others through complex, though often brief, negotiations in which the
relevant information and technical knowledge possessed by them
would have to be declared, and that possessed by others ascertained.

The sheer difficulty of contriving the correct social stance and the
effective social manner for use in different negotiations, the embarrassment of having so to contrive, and the personal affront attached to
failure to achieve one's ends by these means, induced in managers a
nervous preoccupation with the hazards of social navigation in the
structure and with the relative validity of their own claims to authority,
information, and technical expertise.

'Normally', said a departmental manager, 'management has a sort of
family tree showing who is responsible for what, and what he is responsible for. It's a pity there's nothing like that here. It's rather difficult
not knowing; there's a lot of trouble caused by this—you get an assistant to a manager who acts as though he were an assistant manager, a
very different thing.' Another man, a product engineer, said 'One of
the troubles here is that nobody is very clear about his title or status or
even his function.' A foreman, explaining his relationship with senior
managers, said of one, 'It's generally gathered, from seeing T. standing
about looking at the roof when something is being done to it and looking over machines, that he's in charge of plant and buildings.' The
same foreman, discussing his own job, said that when he had first been
promoted he had been told nothing of his duties and functions. 'Of
course, nobody knows what his job is in here. When I was made foreman I was told to get on with the job—was just told "You'll start in on
Monday", so I came in and started in. That was really all that was said.'

The disruptive effects of this preoccupation were countered by a
general awareness of the common purpose of the concern's attitudes.
While this awareness was sporadic and partial for many members of the
firm, it was an essential factor in, for example, the ability of the

'product engineers' to perform their tasks, dependent as they were on the co-operation of persons and groups who carried on the basic interpretative processes of the concern. Indeed, discussion of the common purposes of the organization featured largely in the conversation of cabals and extra-mural groups existing among managers.

ORGANIZATION IN THE LEAST PREDICTABLE CONDITIONS

An even more important part was played by common beliefs and a sense of common purpose in the limiting case of rapidly changing conditions we encountered; a concern recently created to develop electronic equipment and components for the commercial market. While a hierarchy of management may certainly be said to have existed, positions in it were defined almost entirely in terms of technical qualifications. The conversion of this structure of technical expertise into a concern with commercial tasks required a continuous process of self education. There were two major aspects of this process. First, individual tasks in the concern were defined almost exclusively as a consequence of interaction with superiors, colleagues, and subordinates; there was no specification by the head of the concern. Secondly, this continuing definition—and redefinition—of structure depended for its success on effective communication. At the end of a discussion with senior members of staff, it was explicitly acknowledged that the organizational problems of the enterprise turned almost entirely on finding the right code of conduct which would make for effective communication—to avoid occasions, as one head of a laboratory put it, 'when I'm explaining a point to a chap and he says "Yes, yes" and I'm not at all sure whether he's caught on.'

A description of this particular organization at the third meeting* elicited the observation that while it might be interesting to throw a number of people together and wait for an organization to emerge, the stringencies of economics usually prohibited one from doing so. One

* References to the 'first', 'second', and 'third' meetings are to (1) The meeting held in the University of Edinburgh on 13 January 1956 of heads and chief laboratory engineers of the firms which participated in the Scottish study and officials of the Scottish Council to discuss a preliminary statement of results of the study. (2) The meeting of heads of English firms and ministry officials held in the Ministry of Supply to discuss the findings of the Scottish study, on 11 November, 1957. (3) The meeting of heads of firms and Government officials to discuss the report on the English study, held at the Department of Scientific and Industrial Research on 15 July, 1958.

answer to this might be that the concern was the creation of a successful and very economically-minded radio manufacturing company. But this is not a very conclusive answer. The point which the recital of all the cases in this chapter should by now have made clear is that this last concern possessed an organization as specific to its needs and as economical, in terms of the task it had to fulfil, as the first.

CHAPTER 6

Mechanistic and Organic Systems
of Management

The last chapter contained instances of the empirical evidence which has brought us to regard the system of management as a dependent variable. In this chapter we shall try to review the more general considerations which support the view that the effective organization of industrial resources, even when considered in its rational aspects alone, does not approximate to one ideal type of management system, but alters in important respects in conformity with changes in extrinsic factors. These extrinsic factors are all, in our view, identifiable as different rates of technical or market change.* By change we mean the appearance of novelties: i.e., new scientific discoveries or technical inventions, and requirements for products of a kind not previously available or demanded.

There are other 'independent variables' which directly affect the form taken by any management system (although, even conceptually, their independence from each other as well as from the management system, is not to be insisted upon; causal relationships in this, as in other social fields, are not one-way affairs). So far as our own research goes, and the published results of other work, we see two such other dimensions: (i) the relative strength of individual commitments to political and status-gaining ends, and (ii) the relative capacity of the directors of a concern to 'lead'—i.e., to interpret the requirements of the external situation and to prescribe the extent of the personal commitments of individuals to the purposes and activities of the working organization. These are dealt with, on the basis of our own experience, in later chapters, but some preliminary observations on each of these two variables

* The important dimension of scale is dealt with subsequently (pp. 103-6).

are necessary here in order to delimit the area of discourse of the current chapter—the 'working organization'.

THE WORKING ORGANIZATION AND PRIVATE COMMITMENTS

An industrial concern exists in order to carry out a specific task. To exist at all, a concern employs a number of people. These are assigned to specific bits of the total task, which is split up according to traditional or rational principles of the division of labour, and according to the technological equipment available. Co-operation between the several members of a concern is achieved by organization. In setting up an organization, the concern confers and defines, for each member, certain rights to control the activity of others (and of himself) and to receive information, and certain obligations to accept control and transmit information. The way in which a concern confers and defines rights and obligations of this kind constitutes the management system. The form of the system varies with the nature of the concern's task.

This task may remain stable in its most important respects, or it may change. The degree of stability or rate of change calls for different systems by which the activities of the concern are controlled, by which information is conveyed throughout the organization, and by which decisions and actions are authorized.

The members of a concern are recruited to be used, by agreement, as resources to achieve its ends. The activities which are directed in this way, *and* the management system in operation, together form the working organization of the concern.

But the men and women it employs bring in with them other, private purposes of their own. To an extent which varies a great deal from person to person, these purposes may be achieved partly by the return they get from the contract with the employing concern to allow themselves, their physical and mental capacities, to be used as resources. But men and women do not ordinarily yield themselves wholly to use as resources by others; indeed, to do so infringes the human purpose of controlling the situation confronting the individual. In every organized working community, therefore (except those for which religious zeal, political enthusiasm, or some other dedication of the self identifies the personal ends of its members with those they believe to be pursued by the others or by their leader), individuals seek to realize other purposes than those they recognize as the organization's.

Men can attain few of their proximate or ultimate ends by means of their own efforts, unrelated to the conduct of anyone else whatsoever. So that, in addition to the organizational structure of the concern itself, other organizational structures are present through which individuals attempt to realize ends other than those of the concern as such. This private organization is usually called the 'informal structure', to distinguish it from the 'formal structure' of the management system.

The distinction was first made at workshop level, in the Bank Wiring Room study at the Hawthorne plant,[32] which was largely devoted to explaining the social determinants of restriction of output* traditionally regarded as a form of individual delinquency, despite F. W. Taylor's observations on the 'systematic' nature of 'soldiering', or shirking.† In seeking to maintain itself and to pursue its private ends, dictated by a 'logic of sentiments' as against the 'logic of efficiency' of the working organization, the informal group of operatives made use of ostracism, ridicule, and physical punishment ('binging') in order to control the conduct of its individual members. But 'the chief mechanism by which they attempted to control . . . outsiders, supervisors, and inspectors, who stood in a position of being able to interfere in their affairs . . . was that of daywork allowance claims'—i.e., part of the regulatory machinery of the working organization itself, instituted in order to control them. One inspector, for example, 'refused to be assimilated and they helped to bring about his removal by charging him with excessive amounts of daywork. This was the most effective device at their command. Interestingly enough, it was a device provided them by their wage incentive plan. The mechanism by which they sought to protect themselves from management was the maintenance of uniform output records, which could be accomplished by reporting more or less output than they produced and by claiming daywork.'[32]††

---

* 'The great variety of activities ordinarily labelled 'restriction of output' represents attempts at social control and discipline and as such are important integrating processes.'([32], p. 523).

† 'It is well within the mark to state that in nineteen out of twenty industrial establishments the workmen believe it to be directly against their interests to give their employers their best initiatiative, and that instead of working hard to do the largest possible amount of work and the best quality of work for their employers, they deliberately work as slowly as they dare while they at the same time try to make those over them believe that they are working fast.'[33] (p. 33).

†† Another example of the manipulation of an official regulatory system in the private interests of a group of operatives is the adjustment of machine starting and stopping times so as to provide for unofficial breaks[34] (p. 148).

From this beginning, more and more interest has become focused on the conduct, the relationships, the sentiments and beliefs, the affiliations and self-identifications of members of organizations which are irrelevant to, or even incompatible with, the 'formal structure' of the organization when this is conceived as an instrument for the exploitation as resources of its employees, managers as well as operatives. As Gouldner puts it, 'It is obvious that all people in organizations have a variety of "latent social identities"—that is, identities which are not culturally prescribed as relevant to or within rational organizations— and that these do intrude upon and influence organizational behaviour.'[35] The private organizations by which individuals attain their own ends came to have, for some writers, more significance than the working organization itself, which becomes part of the environment the individual seeks to control:

'All formal organizations are moulded by forces tangential to their rationally ordered structures and stated goals. Every formal organization, trade union, political party, army, corporation, etc.— attempts to mobilize human and technical resources as means for the achievements of its ends. However, the individuals within the system tend to resist being treated as means. They interact as wholes, bringing to bear their own special problems and purposes. . . . It follows that there will develop an informal structure within the organization which will reflect the spontaneous efforts of individuals and sub-groups to control the conditions of their existence. . . . It is to these informal relations and structures that the attention of the sociologist will be primarily directed. He will look upon the formal structure, e.g., the official chain of command, as the special environment within and in relation to which the informal structure is built.' ([36] pp. 250-1.)

The presence in a concern of such informal structures may exert a decisive influence over the efficiency of the working organization, and particularly over the degree of appropriateness the type of management system has to the external situation, whether stable or changing, which confronts the concern. In pursuing these private purposes which are irrelevant to the working organization, individuals affiliate themselves to groups, and seek to bind others in association. They acquire commitments. These commitments may persist in the face of an express need for the working organization to adapt itself to new circumstances. Further, commitments involve some surrender of personal autonomy, not, this time, in a bargain with employers and in return for money and other benefits, but in the hope of further material or non-material rewards or in order to avoid discomfort, embarrassment, or loss (perhaps of self-esteem). Commitments to others involve loss of autonomy in that the right to spontaneous, divergent action outside the group is surrendered in respect of the objectives the private combination has been constituted to attain. Voluntarily, or under such indirect restraints and controls as fear of ostracism, ridicule, or damaging criticism through gossip exchanges, individuals submit to their use as resources in the pursuit of private ends tacitly or explicitly formulated by groups. Even the leading figures in private combinations are bound by such controls, as Whyte[37] and, following him, Homans[38] have demonstrated in another setting.

The activities within a concern of private combinations and individuals in pursuit of ends distinct from those of the working organization account for the distortions and frustrations recurrently or chronically experienced by working organizations. Instances in detail of this effect are given in the account of the Bank Wiring Room study already mentioned; Selznick has classified the ways in which such private commitments give rise to unanticipated consequences for the formal organization [36] (p. 255).

In some ways the formal constitution of a concern itself promotes commitments to private ends which may become out of keeping with the purposes of the working organization. Barnard, in his essay 'Functions and Pathology of Status Systems',[39] suggests that since it is impossible to forecast future conditions correctly, and since the members of any concern in our society must be ordered according to some systematic and enduring scale or rank, 'it is necessary to employ and often elaborate a system of status whose inherent tendency is to become unbalanced, rigid and unjust'. The attachment of a person to the position he holds, with all its rights, privileges, and powers, may become an autonomous commitment, enduring after the need has arisen for some change in the character of the position or its location in the structure. Similarly, the desire to defend or assert the rights, privileges, and powers of a position may reduce the effectiveness of communication; subordinates may take it upon themselves to treat instructions and decisions from superiors merely as information for them to use, along with information from other sources, in order to arrive at their own decisions.[40]

The presence of and interplay with each other, and with the working organization, of a multiplicity of commitments, each realized in overt or covert activities, gives rise to a natural social system—to the 'informal organization' of the concern. In analysing these activities, however, we have found it helpful to distinguish two kinds of system —the political system and the status structure. Both systems affect the rational adaptation and exclusive devotion of the working organization to its task, and so introduce a second variable, other than that of technical and commercial change, affecting the form of the management system.

THE WORKING ORGANIZATION AND ITS DIRECTION

So far we have treated the concern as an impersonal entity. There is, however, a central source of visible power, so far as the working organization is concerned, in the Chairman of the Board or the Managing Director, or in the Divisional or the Factory Manager, according to the constitution of the concern. And the extent to which the two major variables already described operate on the form of the management system is largely a matter determined by him, whether he proceeds by conscious study and planning or by intuitive perception and sensitive steering, whether positively by instruction, persuasion and approval, or negatively in permissive tolerance or careless ignorance.

He, in fact, whether alone or in association with other members of the concern, directs the use to which the combined resources of people and materials and equipment of the concern are to be put. It is for him to decide the nature of the task to which the concern is being applied, and, in particular, to gauge the rate of change in the conditions of that task. Once the situation is appreciated the resources of the concern can then be strategically disposed in a management system. This function of the head of the concern, as Selznick has pointed out in general terms, gives his role a qualitative difference from that of subordinate managers. 'Efficiency as an operating ideal presumes that goals are settled and that the main resources and methods for achieving them are available. . . . Leadership goes beyond efficiency (1) when it sets the basic mission of the organization, and (2) when it creates a social organism capable of fulfilling that mission.'[41]

The echo of Mary Parker Follett in this specification for the industrial boss is struck from the notion of leadership, at least at the top of the working organization, as 'emergent', to use her word—creative of new purposes and goals.[42] In our terms, leadership at the top, or 'direction', involves constant preoccupation with the technical and commercial parameters of the situation in which the concern has to operate, and with the adjustment of the internal system to that external situation. It is a different kind of activity from that required in subordinate positions in the management system.

There is a second function of direction, discharged like the first either positively or by default. We have distinguished between the commitment of an individual to the working organization and his commitments to private ends, his own or his associates', which are irrelevant to or incompatible with those of the working organization. In both kinds of commitment the person yields himself in overt and active, or in tacit and compliant ways, as a resource to be exploited by others (in return, of course, for contractual, or promised, or speculative benefits). It is the task of the managing director (sc. chairman, factory manager, etc.) to ensure that the bounds of commitment to the working organization are set wide enough—that the resources provided by the individual for use by the working organization are sufficient—to achieve its purposes. This involves a complex social process by which the director of a concern specifies the requirements of the working organization vis-à-vis other private commitments. He 'defines the work situation', displaying in his own actions and expecting in others' (a) the span of considerations, technical, commercial, humane, politic, sentimental, and so forth, which are admissible to decisions within the working organization; and (b) the demands of the working organization for commitment, effort, and self-involvement which the individual should regard as feasible, and should attempt to meet.

These two directive functions each correspond to one of the other two major variables: (i) the rate of change in the external situation, and (ii) the relative strength of the pursuit of self-interest by members of the concern as against their commitment to the working organization. The variable character of this third element lies in the capacity of the director to fit the management system to its task and to define adequately the work situation of himself and his subordinates.

We shall endeavour to trace the operation of these two major variables, the strength of the private organization and the capability of direction, in succeeding chapters. We have now, however, to proceed to the more precise description of the character of the system on which these variables act and the appropriate forms it assumes in accordance with alteration in the first variable of stability—change.

MODELS OF THE WORKING ORGANIZATION

As industrial enterprises grew in size and as the variety of individual tasks increased with the division of labour and the development of technology, the consequential task was created of directing, co-ordinating, and monitoring the activities of large numbers of men, women, and children. The first spontaneous attempts at comparatively large-scale organization used a system of sub-contracting. Bendix, in his study of industrial management in England during the Industrial Revolution,[43] cites evidence that this system prevailed in iron-making and cutlery, in engineering, building, textiles, pin-manufacture, clothing, match-industry, boot- and shoe-making, printing, paper, mining, and railways, and remarks 'It was obviously up to these subcontractors to deal with their "underhands", whom they recruited, employed, trained, supervised, disciplined, paid and fired.' (p. 53).*

The incorporation of the sub-contractor (and his 'underhands') into the concern itself, first as agent and then as foreman, gradually pruned away many of these functions from him. For most of the nineteenth century, however, the scale of enterprise was such that what functions were removed from the salaried agent or overseer reverted to the owner-manager himself. It is the last hundred years that have seen the absolute and relative growth of the salaried officials of industry. Bendix [43] (p. 214,) has worked out estimates for the period c. 1900–c. 1950 for Britain, France, Germany, Sweden, and the United States. According to his findings, the proportion of 'administrative employees' to 'production employees' rose in Britain from 8.6 per cent in 1907 to 20 per cent in 1948. In Sweden, where this proportion grew most rapidly, it rose from 6.6 per cent in 1915 to 21 per cent in 1950.

We have earlier (pp. 34–5) remarked upon some of the other implications of this phase in the development of the modern industrial enterprise, when typically the small-scale, short-lived concern gave place to the large-scale, enduring corporation. Increase in scale had two aspects: sheer growth in numbers employed, and increase in productivity. Both are equally familiar developments, and both depended largely on the expansion of the consumption of standard products which enabled

* See also Clapham,[44] Vol. II, pp. 124–33.

industrialists to break down manufacturing processes into small individual cycles of activity which could be converted into machine processes
or semi-skilled routines. The whole system depended on holding demand for the same product steady enough and on technical development being slow enough for large-lot production. The history of the
motor-car industry is characteristic of the general industrial trend.

The standardization (through publicity and through price reductions)
of consumer demand enabled the major industries to maintain relatively
stable technical and commercial conditions. Under such conditions not
only did concerns grow in size, not only could manufacturing processes
be routinized, mechanized, and quickened, but the task of ensuring co-
operation and of co-ordination, of planning and monitoring, could also
be broken down into routines and inculcated as specialized management tasks.

The growth in the numbers of industrial administrative officials, or
managers, reflects the growth of organizational structures. Production
department managers, sales managers, accountants, inspectors, and the
rest, emerged as specialized parts of the general management function as
industrial concerns increased in size. Their jobs were created, in fact, out
of the general manager's either directly, or at one or two removes. This
gives them and the whole social structure which contains their newly
created roles its hierarchic character. It is indeed—what one would
expect to emerge from the spontaneous sub-contracting phase of
management if history followed set patterns—a quasi-feudal structure.
All rights and powers at every level derive from the immediate superior; fealty, or 'responsibility', is owed to him; all benefits are 'as if'
dispensed by him. The feudal bond is more easily and more often
broken than in feudal polities, but loyalty to the concern, to employers,
is regarded not only as proper, but as essential to the preservation of the
system. Chester Barnard makes this point with the utmost clarity and
emphasis: 'The most important single contribution required of the
executive, certainly the most universal qualification, is loyalty, domination by the organization personality.'[45] (p. 220). And, as A. W.
Gouldner has said, 'Much of W. H. Whyte's recent study of "organization man" is a discussion of the efforts by industry to attach managerial loyalty to the corporation'. [35] (p. 416).

In drawing a contrast between the spontaneity of the first phase of
industrial management and the highly structured formalistic nature of
the second, we do not imply in either case that their adoption by

shrewd and competent and successful business men was anything other than a reasonable solution for their immediate problem. Each was—and is—appropriate to specific sets of circumstances, circumstances which include the knowledge available of the working of administrative systems. At the time of the creation of formalized management systems, this knowledge was limited largely to the experience of large, stable bureaucracies—civil service establishments and ecclesiastical hierarchies—and to military forces and ships' companies. Industry, when the time came, inevitably adopted bureaucracy as its management system because, during this phase, stability was a basic presumption of growth.

The formal organization of industrial management along bureaucratic lines coupled with the concurrent growth of national armies and governmental administrations, especially in Western Europe, suggested to sociologists that 'bureaucratization' was as intrinsic to the character of modern society as was scientific and technological progress. For Weber, the founder of the study of bureaucracy, it exhibited the same feature of rational thought applied to the social environment as does technology in the case of the physical environment.*

Bureaucracy, then, stands as the 'formal organization' of industrial concerns. The formulation given by Weber is a generalized description of the 'ideal type' of bureaucracy—i.e., a synthetic model composed of what are understood in society at large to be the distinguishing features of actual bureaucratic organizations, military, ecclesiastical, governmental, industrial, etc. These distinctive characteristics are:

(i) The organization operates according to a body of laws or rules, which are consistent and have normally been intentionally established.

* Bureaucracy is often said to have played the part in Weber's sociology that social class did in Marx's. But both men were preoccupied, beyond those notions, with the idea of 'alienation'—the moral, intellectual, and social constraints exercised over men's natural, instinctual inclinations by the immense apparatus of the social order. 'Total bureaucracy', the emergent ideal of the absolute State, was feared by Weber as it was later accepted, even welcomed, by Jünger. But it is to be distinguished from the monolithic 'totalitarian state' in which the civil service and army are subordinate to the Party and its leader. In this latter case bureaucracy, as Burin has pointed out, instead of becoming the dominant element in the life of the nation, because its dominance is presumed to be essential to the nation's survival, is an instrument by which the nation is controlled in the interests of the Party's survival.[46] For Jünger's exposition of organization in every sector of life as necessary for the only purposes acceptable to modern society—or left for it to believe in—see Der Arbeiter.[47]

(ii) Every official is subject to an impersonal order by which he guides his actions. In turn his instructions have authority only insofar as they conform with this generally understood body of rules; obedience is due to his office, not to him as an individual.

(iii) Each incumbent of an office has a specified sphere of competence, with obligations, authority, and powers to compel obedience strictly defined.

(iv) The organization of offices follows the principle of hierarchy; that is, each lower office is under the control and supervision of a higher one.

(v) The supreme head of the organization, and only he, occupies his position by appropriation, by election, or by being designated as successor. Other offices are filled, in principle, by free selection, and candidates are selected on the basis of 'technical' qualifications. They are appointed, not elected.

(vi) The system also serves as a career ladder. There is promotion according to seniority or achievement. Promotion is dependent on the judgement of superiors.

(vii) The official who, in principle, is excluded from any ownership rights in the concern, or in his position, is subject to discipline and control in the conduct of his office.[48]

These general principles underlie every subsequent definition given to formal organization in industry; they can even be read, in garbled form, in Henri Fayol's principles of management,[49] although it is extremely unlikely that Weber's work was known to him. Fayol's definitions of management, organization, and so on amount to no more than a thesaurus of synonyms, but the visible symbol of formal organization—the organization chart—originates with him and remains, along with the 'organization manual' of job descriptions, the chief instrument of industrial organization.[50, 51] Industrial consultants, the disciples of Fayol and of F. W. Taylor (whose discrepant views about the necessity of unified command and functional specialization still survive actively in the dilemma of the line and staff organizational structure) have refined and developed the bureaucratic conception of organization in keeping with the growth in technical complexity and scale of twentieth-century industry.

In discussing management systems, social scientists have usually followed one of two paths. They have either accepted the organization chart and manual conception as the 'formal organization'—an imposed system of control, information, and authority to which seniors try to get their subordinates to conform—or have harked back to Weber's ideal type of bureaucratic structure and proposed this as a rationalistic interpretation of the working organization of a concern.

The first case seems to entail the notion of a concern as two mutually opposed social systems. The Manichean world of the Hawthorne studies (the chapter of *Management and the Worker* dealing with these matters is headed 'Formal *versus* Informal Organization') has been left behind, but it has been succeeded by a crudely Freudian dualism, with formal organization in the role of consciousness, and the concealed or repressed informal organization up to all kinds of mischief.

'The *theory* of formal organization is, in itself, quite simple. It holds that throughout the organization there is a strict definition of authority and responsibility. Similarly, there is an equally precise definition of the function of every department. . . . In addition to the described attributes of formal organization, there are several implied assumptions. The first of these is that formal organization is necessary to achieve organizational goals. It is necessary because it is by nature impersonal, logical, and efficient. An organization can function best when individual idiosyncrasies, sentiments, and prejudices do not interfere with official activities. . . .

'The second assumption of formal organization is that it is the only organization. . . .

'Although formal organization is designed to subject production to logical planning, things never seem to go "according to plan". This is evidenced by the many "problems" managers encounter. They find that no matter how carefully they organize, despite the concern in anticipating problems, unanticipated ones always arise. For these eventualities formal organization offers little guidance because it is created as a guidepost for the routine, the typical, and the foreseeable.'[52] (pp. 159-60)

For an explanation of these unanticipated consequences, the authors prescribe a study of the 'informal organization'.

The second view construes the bureaucratic structure as one of two possible 'models' of the working organization. Gouldner,[35] (p. 405), suggests that the bureaucratic model is a consequence of a rationalistic

---

* See also Roethlisberger's *Management and Morale*,[53] ch. V, 'A Disinterested Observer Looks at Industry'; Moore's *Industrial Relations and the Social Order*,[54] ch. VI 'Blueprint Organization', and ch. XV 'Informal Organization of Workers'.

view, in contrast with the 'natural-system' model (such as that presented by Selznick in his *TVA and The Grass Roots*) which arises from the realization that the commercial or administrative 'goals of the system as a whole are but one of several important needs to which the organization is oriented. Its component structures are seen as emergent institutions, which can be understood only in relation to the diverse needs of the total system. The organization, according to this model, strives to survive and to maintain its equilibrium, and this striving may persist even after its explicitly held goals have been successfully attained. This strain towards survival may even on occasion lead to the neglect or distortion of the organization's goals. Whatever the plans of their creators, organizations, say the natural system theorists, become ends in themselves and possess their own distinctive needs which have to be satisfied'. Waldo has even identified four different models, each interpreting the working organization from a different point of view: the 'machine model' conceived in terms of efficient procedures; the 'business model' with activities interpreted in terms of their profitability; the 'organic model', which presents the relationships of the concern and its members with the total environment; and the 'pure system' model, emphasizing the nature of any organization as a system with special systemic needs.[55] These four models are, however, easily reducible to Gouldner's two when structure and function are considered together: the 'machine model' is, after all, a structural representation usually associated with the manipulation of the resources of a concern in the interests of profit (the 'business' model); and the 'pure system' model, as Gouldner has seen, is merely the 'organic model' seen from the point of view of the total needs which the organic system is created and maintained to satisfy.

What all these kinds of view have in common is the assumption that they are concerned with the same thing, that all working organizations are analysable in one or other set of terms, the choice depending not on the difference between working organizations but on the different standpoints of the writers. Variations in working organizations themselves are attributed to departures from normality, or from the overt purpose of the concern, which may be read as inefficiency (Fayol, etc.), irrationality (Weber, etc.), pathology (Mayo, Barnard, etc.), or the superior strength or inflexibility of 'informal' institutions (Selznick, etc.).

In the last few years, however, there have emerged some attempts at a synthetic appreciation of the concern which will accept the fact that it is both a bureaucratic institution with a specific social purpose to fulfil *and* a community of people with distinct purposes and institutional forms. One empirical study aimed in this direction is Gouldner's own study of the 'succession crisis' in a gypsum mine and factory, when an 'outside man' took over the management.[56] The particular form taken by any working organization, he suggested, was the result of the predominance of one or two distinct bureaucratic patterns, either of which took its shape from the intentions and aspirations of chief managers and from the responses and counter systems evolved by subordinates. A 'representative bureaucracy' is one in which the organizational rules are initiated and maintained by a majority of all the individuals, managers and workers, on the grounds that the rules are means to ends they all, in some measure, desire. (A 'mock bureaucracy' may also exist as a tacit, collusive system in which the formal rules are disobeyed and remain unenforced.) The second pattern is 'punishment-centred', a system by which rules are enforced on the grounds on the authority vested in superior office, subordinates are ordered to do things divergent from their own ends, and resistance from subordinates counterbalances the stress on obedience as an end in itself.[35] (p. 403).

The point of the study lies not so much in the formulation of varieties of bureaucracy as in the way in which the actual conduct of a bureaucracy of either kind is displayed as a matter of 'informal cooperation and spontaneous reciprocity ... the latent function of bureaucratic rules (being) to provide a managerial indulgency, in the form of withholding applications of the rules' (*ibid*). In this sense, Gouldner's variable faintly echoes our third independent variable, 'direction', which has been adumbrated earlier in this chapter, and to which we shall revert in Part III.

The second attempt at a synthesis is associated, so far as sociology is concerned, with the name of H. A. Simon and with the development of organization theory.

The main preoccupation of organization theory is again with the internal structure of organizations and their efficiency, but students have directed their attention especially to what had previously been accepted as a datum—the rationality of the formal organization. It is, as Eisenstadt says, 'concerned with the conditions which make for maximum rational behaviour, calculation, and performance within a given structural organizational setting, or, conversely, the extent to which various structures and organizational factors limit rational calculation and efficiency'.[57]

In itself an eclectic branch of sociology, drawing on classical economic theory, communication theory, the study of small groups, and industrial sociology, organization theory has become associated with a number of other parallel developments which are relevant to the study of working organizations. Further, it is identified with a much wider movement in the social sciences at large.

ORGANIZING FOR CHANGE

In many of the social sciences, one focal point of interest during the past decade has been the attempt, increasingly determined, to replace or supplement the static theoretical models of their textbooks with dynamic models. The data of economics, of psychology, and of sociology are largely events and transactions, although these happenings may be often, even largely, repetitive, or routine, or customary. These studies are also concerned with the conduct of actual people, although the existence of a large number of abstract terms in common usage and technical language allows us to ignore this for quite long stretches of time. Awareness of this has led to a number of assaults on accepted theories, which presume an order in equilibrium, or of attempts to square the accepted theories with the facts of life.

Thus E. R. Leach: 'The model systems of earlier anthropologists ... have a common element in that they are conceived of as stable fixed systems, they are ideal types. Modern social anthropologists usually operate on a much more modest scale in much greater detail, but their "societies" are still, I maintain, largely model systems, the stability of which is an hypothesis not an established fact. . . . My own view is that equilibrium theory in social anthropology was once justified but that it now needs drastic modification. . . . We must recognize that few, if any, of the societies which a modern field worker can study show any marked tendency towards stability.'[58]*

* The innovation which Leach proposes, after this bold manifesto, is slightly disappointing: 'In practical field work situations the anthropologist must always treat the material of observation *as if* it were part of an overall equilibrium, otherwise description becomes almost impossible. All that I am asking is that the fictional nature of this equilibrium should be frankly recognized' (*ibid*, p. 285). But at least the criticism is downright.

F. H. Allport, more recently, has produced a detailed and lengthy critical review of a number of theories of perception. At the end, he suggests that the inadequacies he finds in all of them are reducible for the most part to the distortions inevitable in the conversion of streams of events to timeless theoretical structures. Full explanation demands that events be allowed for in the statement of 'laws governing the material situation'. 'Events are necessary as a kind of "framework" within which dimensional or quantitative laws appear.'[59]

Both in psychology and sociology, however, there seems to be little more at present than some dissatisfaction with the existing state of affairs. Social psychologists have been less inhibited, at least in America; there is a lively interest in experiments on co-operation and competition,[60] gambling,[61] and the performance of groups with work to do or problems to solve.[62, 63, 64] Undoubtedly, however, it is economists who have formulated most explicitly (and modestly) the criticisms of static theories and have produced the most fertile and rapidly exploited ideas on which to construct an 'economic dynamics'.

The critical attitude towards the orthodox tradition is much the same as in other social sciences; the editors of the *Symposium on Uncertainty and Business Decisions* remark that such innovators 'have now come to a parting of the ways: either they must be content with mechanical systems which ignore the complexity of men's thoughts and motives and treat past and future as essentially indistinguishable from each other; or else they must try to understand how men decide upon their courses of action when they cannot feel sure, even within wide limits, what the outcome of any course will be.'[65] But the latter part of the statement, the second alternative, is much more than a pious hope or a programme. In particular, the development of conceptual models which will take account of the empirical situation of people confronted with decisions to make has proved of exceptional interest for sociologists. This is not only because of the intrinsic interest of what G. L. S. Shackle has made of the notions of belief and potential surprise (uncertainty) in developing his model of decision-making, but also because this line of theorizing has concurred, and finally joined up, with the gradual concentration of interest among sociologists on the 'decision-making' function as the central element of the role of business manager[30] (especially Introduction to 2nd ed.),[45] (especially Chs. XII and XIV).

Put simply, the general thesis expounded by Shackle[66, 67, 68] and accepted by those who have since joined forces and issues with him, runs thus: the course actually followed by a firm, or by an isolatable section of a firm, or, in the last resort, by the individual business man, is set by choices between two or more possible courses of action. Business decisions hinge on (a) the relative gain or loss attached to the various outcomes, and (b) the strength or weakness of the belief accorded to each expectation. The attractiveness of courses of action will be a function of the several combinations of profitability (or its reverse) and feasibility (or unlikelihood) which they represent.

The controversial part of this conception, and that which has proved the most fertile ground for later development and speculation, is the reconstruction of 'feasibility' of expectations in terms other than those of orthodox notions of probability which is 'measured by the relative frequency of favourable cases among all possible cases, provided these are equally possible',[69] or, in other words supplied by W. B. Gallie, is 'the ratio between one particular sort of result to (sic) all other allegedly possible results in a theoretically repeatable experiment'.[70] The innovation here lies in deserting the assumption that business (and most other) decisions are based on any frequency-ratio. 'Mathematics has shown in probability theory how to derive knowledge concerning aggregates of future events from aggregates of past events, but this technique is irrelevant to personally unique and crucial occasions of decisions by an individual.'[71] We may observe that Toulmin has also argued, more generally, that statistical calculations of probability are in fact no more than refinements of the guarded conclusion or the qualified conclusion of common usage and not statements of a totally different logical status.

As usual with new departures initiated by a sense of the inadequacy of existing theory, great difficulty has been experienced in determining the boundaries of the constellation of meanings to be attached to expectation in this new guise. The focal concept is uncertainty—the ignorance of the person who is confronted with a choice about the future in general, and in particular about the outcomes which may follow any of his possible lines of action. Since he must choose, if he is to remain operative (as a business man or any other agent), he acts in accordance with his belief about the future and the specific possibilities. These possibilities will always be differentiated in his mind according to the degrees of belief with which they are credited.

DISCONTINUOUS VARIATION IN BELIEF

At this point, decision-making discutants tend to divide into two groups. The first appear to be concerned with the construction of a calculus out of 'degrees of belief'. To do this, Shackle himself invokes the notion of 'potential surprise' as an indication of the degree of belief *in reverse*, it being nonsense to talk of 'positive belief' in the case of a number of possibilities of which all a man can say is that he is not confident that they won't happen—i.e. he has to take them into account. He goes on 'I have called belief a feeling, but I think it may be different in kind, or perhaps more complex, compared with such feelings as pleasure, tension, and excitement. It might even convey something of the truth if we said that belief is an attitude or (temporary) configuration of the mind, rather than an activity that can in itself have various degrees of intensity. If so, we need some symptom whose own differences of intensity will reflect these differences of attitude between states of high and low disbelief. Now when something occurs whose possibility we had disbelieved in, we experience surprise.' Shackle's 'potential surprise' is this feeling projected into the conditional future—how much more surprised one would be if this, rather than that, occurred, and so on, through all the possibilities confronting one at this moment: 'My proposal is to regard the surprise with which we now imagine the occurrence of some specified event at a future date would occasion in us if our knowledge and inferences remained unchanged from now until then, as the representative of our present disbelief in that event, as what makes that disbelief an effective factor in our decision-making.'[68] (p. 30).

It is, however, the second area of discussion which is important in the present context. This begins by questioning the existence in fact of a continuum along which expectations may be ranked in order not of their utility or value (or probability). Thus C. F. Carter: 'Professor Shackle . . . provides for a scale of potential surprise, personal to an individual; the essential condition for the existence of such a scale being an ability to compare differences in potential surprise. I am not sure that this is very plausible; at least, I do not find it a reasonable assumption to make about degrees of belief. . . . It is a logical impossibility to devise a means of compounding two pure rankings (or one ranking and a numerical index) to make another unique ranking. . . . To fight an act of aggression may do both more good and more evil than to give in to it; and the agony of deciding what to do in such a situation is surely due to the existence of contrary rankings of preference side by side, and incapable of being given a numerical measure.'[73]

For a solution, Carter refers to what he suggests happens in practice: outcomes are ranked in broad groups, the 'perfectly possible', the 'just possible', and so on, with no differentiation so far as belief is concerned between the members of each group. This is, in fact, he suggests, what Shackle himself does with his own 'perfectly possible' group of outcomes. Thus, decision results from considering possible courses of action in the light of discontinuous steps of belief. 'This seems to me to be the rational way of thinking. "Course A will be better than Course B if there is inflation; but Course B is wiser if there is to be a change of Government" ... and so on' (*ibid*, p. 56).

The importance of Professor Carter's comments lies in the suggestion, which seems to us well-founded, that the judgements on which decisions are based have to bridge discontinuous and often unrelatable sets of contingencies. Even if this 'field', to use Toulmin's word, remains the same, one often decides that a particular course of action looks most worth while under the circumstances and then a 'But suppose—?' is interjected; with the arrival of a fresh set of considerations, one is confronted with the need to arrive at a decision as to the relevance or degree of belief in the new set of considerations and then decide afresh on a course of action, all with possibly different canons of judgement from what obtained before.

TYPES OF DECISION-MAKING SITUATION

H. A. Simon approaches the theme of discontinuity from a totally different point of departure—what he calls 'the facts of life that we will discover when we make actual empirical studies of the formation and use of expectations in business decision-making.'[74] What transpires is very closely akin to what Shackle has treated in terms of 'seriable' and 'non-seriable' decision-making. 'Seriable' decisions are those which are very frequently repeated, with expectations relating quite specifically to past experience in similar circumstances. Seriable decisions involve virtually no uncertainty and the lowest potential surprise, and are next in order of uncertainty to decisions which are insurable against consequent losses (and are therefore equivalent to 'dead certs'). The discussion of decision-making proper, for Shackle, begins with 'non-seriable' decisions.

Simon suggests that, by common knowledge, a large number of the decisions we ordinarily make, in business and elsewhere, fall within the limits of a 'programme': 'Under certain circumstances when an individual or an organization is confronted with a situation requiring decision, the decision process goes off quickly and smoothly—almost as though no decision were being made at all, but the matter had been decided previously.' By contrast, 'non-programmed' decisions involve 'much stirring about, deliberation, discussion, often vacillation'. The distinction is said to resemble that drawn by other, psychologist, students of decision-making between 'habitual behaviour' and 'genuine decisions' or between 'routine' and 'critical' decisions.

The grading introduced by this kind of division seems at first sight to be different from Carter's discontinuous array of fields of decision. But what Simon is doing amounts to no more than substituting 'doubt' for 'potential surprise'. In programmed decisions the choice made is to some extent a foregone conclusion, although it may be necessary to perform a complicated series of calculations before a single choice is actually made. In fact, it corresponds to our conception of interpreting local and technical information in relation to a specific choice between courses of action. Simon instances driving a car to a destination, which involves almost exclusively programmed decision-making, although a good deal of computation (most of which is done 'unthinkingly') takes place based on information obtained through the eyes. Experience in driving counts, but the more experienced the driver, the nearer the approach to a fully programmed series of decisions—i.e. he knows what to do 'automatically' on every occasion of choice. There is no substantial dissimilarity, says Simon, between this and what happens in practice when a monthly schedule is planned in a factory.

Programmed decision-making is what it is because of the existence of an institutional framework around the individual. 'The pattern of behaviour in a business firm in which this particular decision' (the choice of an aggregate production rate for a factory) 'represents one small detail, may be regarded as a mosaic of such decision-making programmes. . . . So far as any of the programmed decision-making processes is concerned, all the other programmes that surround it are a part of its environment.'

This squares with an earlier thesis of the same writer on 'composite decision' as the process by which an organization influences the decisions of each of its members—supplying these decisions with their premises. 'It should be perfectly apparent that almost no decision made in an organization is the task of a single individual. Even though the final responsibility for taking a particular action rests with some definite person, we shall always find, in studying the manner in which this decision was reached, that its various components can be traced through the formal and informal channels of communication to many individuals who have participated in forming its premises.'[30] (p. 221).*

In non-programmed decisions 'the alternatives of choice are not given in advance, but must be discovered' by a rational process of searching. Not that in non-programmed decisions the chooser is compelled to calculate in terms of expectations and his degree of belief in them; Simon suggests that such searches for alternatives are conditional upon a higher level of aspiration, when satisfaction will have to be sought outside the routine, habitual, programmed courses of decision-making.

We have now placed Simon in a similar position to that established by Carter, although Simon has elaborated the general thesis a good deal. The reconciliation lies in the fact that to arrive at a decision in the circumstances envisaged by Carter, when disparate sets of considerations enter in, requires a search, in Simon's terms, for the wider set of considerations which will provide a frame of reference within which both former sets may be evaluated. Both, it is important to notice, start from a strictly empirical examination of their experience of decision-making.

In place of a reckoning based on a makeshift, fictionally objective, probability distribution of expectations, or a subjective probability distribution, or a potential surprise curve, they suggest that the span of expectations involved in a decision expands stepwise and discontinuously by the interjection or supersession of new sets of considerations, not in the same field as those which apply to what were the safe or familiar expectations.

As Colin Cherry has pointed out, the search for new information implies doubts.[75] As soon as a 'But suppose—?' question has intruded itself into the consideration of a specific set of possible courses of action, the existence of other possibilities is implied. For Carter, these other possibilities may be such that they cannot be compared with the first set by criteria applicable to either, nor may there be a third set of

---

* cf. Mary Parker Follett. 'An order or command is a step in a process, a moment in a movement of interweaving experience. We should guard against thinking this step is a larger part of the whole process than it really is There is all that leads to the order, all that comes afterwards—methods of administration, the watching and recording of results, what flows out of it to make further orders.'[42](pp. 149-150).

criteria in existence available for a rational choice between them. For Simon, these other possibilities represent a departure from an institutional matrix for decisions (within which choosing is a matter of the straightforward computation of familiar information), and a search for new considerations which may prove to be more remote institutions or the foundations of new institutional processes.

A century ago, Herbert Spencer described intelligence as the 'establishment of correspondence between the relations in the organism and relations in the environment', a formulation which can now be treated as axiomatic by neurophysiologists.[76] Ten years ago, a psychologist, G. P. Meredith, represented the brain as a continuously interrelated system or structure-begetting apparatus through which the organism comes to terms with its environment.[77] Later, however, he reduced the emphasis on the 'epistemic' unit of the mind, and has concentrated on the heuristic representations of the environmental situation (epistemic moduli) with which the individual operates.[78] The unity embracing these different moduli, and giving each an individuality specific to the person, is biography—the personal and social experience of the self.

Equipped with this apparatus of representational constructions he puts on the world through which he has lived, the individual is enabled, more or less successfully, to act in accordance with his expectations. There is here a striking similarity between Meredith and Simon—the former writes:

'Genetic constitution and growth processes build the individual into an expectant being on the basis of past events. Given that the environment usually changes very slowly, the constancy of its characteristics makes it a framework of type events which recur with virtually predictable regularity.... Thus life is organized on the basis of a gigantic hypothesis. We may call it "the hypothesis of no surprise" (cf. Simon's suggestion that most decisions are programmed —i.e., determined for us by our continuous apprehension of the framework of social institutions around us).

'But then we encounter the philosophical puzzle that the environment does, in fact, present surprises, and nevertheless life goes on. It even advances. The explanation of survival in a surprising world is that not all surprises are fatal. Meeting a surprise the organism finds its immediate hypothesis unfulfilled. This occasions a state of stress which calls forth a second hypothesis of a somewhat more general character. The organism has an appropriate behaviour-pattern grounded in this second hypothesis.'[78] (p. 75).

The notion of 'continuous expectations' which Meredith regards as a 'basic characteristic of living organisms' is more familiar sociologically as the perpetual call for action put upon the individual,* Faced with these demands for action, involving decisions, the individual is constantly re-creating for himself operational representations of the situations through which he moves. 'Living is a moment-to-moment affair, and the translation of' (others' and our own) 'conduct and events into non-temporal patterns provides us with a kind of chart for the guidance of next action; at any one moment, that is, we need to be concerned simultaneously and systematically with the events, persons, and other objects we believe relevant to our own conduct at that moment.'[80]

Guides for action, epistemic moduli, or institutions are not wholly private to the individual. They are, as Parsons has pointed out in another connexion, functions of interaction between persons. As such, their existence depends on a pre-existing *common culture—that is, a commonly shared system of symbols the meanings of which are understood on both sides with an approximation to agreement'.[81] Non-verbal conduct, as well as objects and language, is involved in such symbol systems.

The sets of patterns of considerations taken into account in decision-making may therefore be regarded as aspects either of the individual person (biographically determined) or of the social context in which a decision is made. Neither will yield, by itself, a comprehensive statement about the framework of belief in which a decision is made. But in working organizations decisions are made either in the presence of others or with the knowledge that they will have to be implemented, or understood, or approved by others. The set of considerations called into relevance on any decision-making occasion has therefore to be one shared with others or acceptable to them.

Our own studies suggest that there are industrial concerns for which non-programmed decision-making is a normal function; indeed, that this kind of activity takes up most management time, and is its most

---

* W. G. Henry in his comment on Meredith's paper suggests that the external situation of the individual is constantly and inherently stressful.[79]

important function. Such firms, in so far as they are successful, have either spontaneously or deliberately worked out a kind of management system which will facilitate non-programmed decision-making. In exploiting human resources in this new direction, such concerns have to rely on the development of a 'common culture', of a dependably constant system of shared beliefs about the common interests of the working community and about the standards and criteria used in it to judge achievement, individual contributions, expertise, and other matters by which a person or a combination of people are evaluated. A system of shared beliefs of this kind is expressed and visible in a code of conduct, a way of dealing with other people. This code of conduct is, in fact, the first sign to the outsider of the presence of a management system appropriate to changing conditions.

MECHANISTIC AND ORGANIC SYSTEMS

We are now at the point at which we may set down the outline of the two management systems which represent for us (see Chap. 5) the two polar extremities of the forms which such systems can take when they are adapted to a specific rate of technical and commercial change. The case we have tried to establish from the literature, as from our research experience exhibited in the last chapter, is that the different forms assumed by a working organization do exist objectively and are not merely interpretations offered by observers of different schools.

Both types represent a 'rational' form of organization, in that they may both, in our experience, be explicitly and deliberately created and maintained to exploit the human resources of a concern in the most efficient manner feasible in the circumstances of the concern. Not surprisingly, however, each exhibits characteristics which have been hitherto associated with different kinds of interpretation. For it is our contention that empirical findings have usually been classified according to sociological ideology rather than according to the functional specificity of the working organization to its task and the conditions confronting it.

We have tried to argue that these are two formally contrasted forms of management system. These we shall call the mechanistic and organic forms.

A *mechanistic* management system is appropriate to stable conditions. It is characterized by:

(a) the specialized differentiation of functional tasks into which the problems and tasks facing the concern as a whole are broken down;

(b) the abstract nature of each individual task, which is pursued with techniques and purposes more or less distinct from those of the concern as a whole; i.e., the functionaries tend to pursue the technical improvement of means, rather than the accomplishment of the ends of the concern;

(c) the reconciliation, for each level in the hierarchy, of these distinct performances by the immediate superiors, who are also, in turn, responsible for seeing that each is relevant in his own special part of the main task.

(d) the precise definition of rights and obligations and technical methods attached to each functional role;

(e) the translation of rights and obligations and methods into the responsibilities of a functional position;

(f) hierarchic structure of control, authority and communication;

(g) a reinforcement of the hierarchic structure by the location of knowledge of actualities exclusively at the top of the hierarchy, where the final reconciliation of distinct tasks and assessment of relevance is made.*

(h) a tendency for interaction between members of the concern to be vertical, i.e., between superior and subordinate;

(i) a tendency for operations and working behaviour to be governed by the instructions and decisions issued by superiors;

(j) insistence on loyalty to the concern and obedience to superiors as a condition of membership;

(k) a greater importance and prestige attaching to internal (local) than to general (cosmopolitan) knowledge, experience, and skill.

* This functional attribute of the head of a concern often takes on a clearly expressive aspect. It is common enough for concerns to instruct all people with whom they deal to address correspondence to the firm (i.e., to its formal head) and for all outgoing letters and orders to be signed by the head of the concern. Similarly, the printed letter heading used by Government departments carries instructions for the replies to be addressed to the Secretary, etc. These instructions are not always taken seriously, either by members of the organization or their correspondents, but in one company this practice was insisted upon and was taken to somewhat unusual lengths; all correspondence was delivered to the managing director, who would thereafter distribute excerpts to members of the staff, synthesizing their replies into the letter of reply which he eventually sent. Telephone communication was also controlled by limiting the numbers of extensions, and by monitoring incoming and outgoing calls.

The *organic* form is appropriate to changing conditions, which give rise constantly to fresh problems and unforeseen requirements for action which cannot be broken down or distributed automatically arising from the functional roles defined within a hierarchic structure. It is characterized by:

(*a*) the contributive nature of special knowledge and experience to the common task of the concern;

(*b*) the 'realistic' nature of the individual task, which is seen as set by the total situation of the concern;

(*c*) the adjustment and continual re-definition of individual tasks through interaction with others;

(*d*) the shedding of 'responsibility' as a limited field of rights, obligations and methods. (Problems may not be posted upwards, downwards or sideways as being someone's else's responsibility);

(*e*) the spread of commitment to the concern beyond any technical definition;

(*f*) a network structure of control, authority, and communication. The sanctions which apply to the individual's conduct in his working role derive more from presumed community of interest with the rest of the working organization in the survival and growth of the firm, and less from a contractual relationship between himself and a non-personal corporation, represented for him by an immediate superior;

(*g*) omniscience no longer imputed to the head of the concern; knowledge about the technical or commercial nature of the here and now task may be located anywhere in the network; this location becoming the *ad hoc* centre of control authority and communication (*cf.* [82]);

(*h*) a lateral rather than a vertical direction of communication through the organization, communication between people of different rank, also, resembling consultation rather than command;

(*i*) a content of communication which consists of information and advice rather than instructions and decisions;[40]

(*j*) commitment to the concern's tasks and to the 'technological ethos' of material progress and expansion is more highly valued than loyalty and obedience;

(*k*) importance and prestige attach to affiliations and expertise valid in the industrial and technical and commercial milieux external to the firm.

One important corollary to be attached to this account is that while organic systems are not hierarchic in the same sense as are mechanistic, they remain stratified. Positions are differentiated according to seniority —i.e., greater expertise. The lead in joint decisions is frequently taken by seniors, but it is an essential presumption of the organic system that the lead, i.e. 'authority', is taken by whoever shows himself most informed and capable, i.e., the 'best authority'. The location of authority is settled by consensus.

A second observation is that the area of commitment to the concern —the extent to which the individual yields himself as a resource to be used by the working organization—is far more extensive in organic than in mechanistic systems. Commitment, in fact, is expected to approach that of the professional scientist to his work, and frequently does. One further consequence of this is that it becomes far less feasible to distinguish 'informal' from 'formal' organization.

Thirdly, the emptying out of significance from the hierarchic command system, by which co-operation is ensured and which serves to monitor the working organization under a mechanistic system, is countered by the development of shared beliefs about the values and goals of the concern. The growth and accretion of institutionalized values, beliefs, and conduct, in the form of commitments, ideology, and manners, around an image of the concern in its industrial and commercial setting make good the loss of formal structure.

Finally, the two forms of system represent a polarity, not a dichotomy; there are, as we have tried to show, intermediate stages between the extremities empirically known to us. Also, the relation of one form to the other is elastic, so that a concern oscillating between relative stability and relative change may also oscillate between the two forms. A concern may (and frequently does) operate with a management system which includes both types.

The organic form, by departing from the familiar clarity and fixity of the hierarchic structure, is often experienced by the individual manager as an uneasy, embarrassed, or chronically anxious quest for knowledge about what he should be doing, or what is expected of him, and similar apprehensiveness about what others are doing. Indeed, as we shall see later, this kind of response is necessary if the organic form of organization is to work effectively. Understandably, such anxiety finds expression in resentment when the apparent confusion besetting him is not explained. In these situations, all managers some of the time, and many managers all the time, yearn for more definition and structure.

On the other hand, some managers recognize a rationale of non-definition, a reasoned basis for the practice of those successful firms in which designation of status, function, and line of responsibility and authority has been vague or even avoided.

The desire for more definition is often in effect a wish to have the limits of one's task more neatly defined—to know what and when one doesn't have to bother about as much as to know what one does have to. It follows that the more definition is given, the more omniscient the management must be, so that no functions are left wholly or partly undischarged, no person is overburdened with undelegated responsibility, or left without the authority to do his job properly. To do this, to have all the separate functions attached to individual roles fitting together and comprehensively, to have communication between persons constantly maintained on a level adequate to the needs of each functional role, requires rules or traditions of behaviour proved over a long time and an equally fixed, stable task. The omniscience which may then be credited to the head of the concern is expressed throughout its body through the lines of command, extending in a clear, explicitly titled hierarchy of officers and subordinates.

The whole mechanistic form is instinct with this twofold principle of definition and dependence which acts as the frame within which action is conceived and carried out. It works, unconsciously, almost in the smallest minutiae of daily activity. 'How late is late?' The answer to this question is not to be found in the rule book, but in the superior. Late is when the boss thinks it is late. Is he the kind of man who thinks 8.00 is the time, and 8.01 is late? Does he think that 8.15 is all right occasionally if it is not a regular thing? Does he think that everyone should be allowed a 5-minutes grace after 8.00 but after that they are late?'[83]

Settling questions about how a person's job is to be done in this way is nevertheless simple, direct, and economical of effort. We shall, in a later chapter, examine more fully the nature of the protection and freedom (in other respects than his job) which this affords the individual.

One other feature of mechanistic organization needs emphasis. It is a necessary condition of its operation that the individual 'works on his own', functionally isolated; he 'knows his job', he is 'responsible for seeing it's done'. He works at a job which is in a sense artificially abstracted from the realities of the situation the concern is dealing with, the accountant 'dealing with the costs side', the works manager 'pushing production', and so on. As this works out in practice, the rest of the organization becomes part of the problem situation the individual has to deal with in order to perform successfully; i.e., difficulties and problems arising from work or information which has been handed over the 'responsibility barrier' between two jobs or departments are regarded as 'really' the responsibility of the person from whom they were received. As a design engineer put in, 'When you get designers handing over designs completely to production, it's "their responsibility" now. And you get tennis games played with the responsibility for anything that goes wrong. What happens is that you're constantly getting unsuspected faults arising from characteristics which you didn't think important in the design. If you get to hear of these through a sales person, or a production person, or somebody to whom the design was handed over to in the dim past, then, instead of being a design problem, it's an annoyance caused by that particular person, who can't do his own job—because you'd thought you were finished with that one, and you're on to something else now.'

When the assumptions of the form of organization make for preoccupation with specialized tasks, the chances of career success, or of greater influence, depend rather on the relative importance which may be attached to each special function by the superior whose task it is to reconcile and control a number of them. And, indeed, to press the claims of one's job or department for a bigger share of the firm's resources is in many cases regarded as a mark of initiative, of effectiveness, and even of 'loyalty to the firm's interests'. The state of affairs thus engendered squares with the role of the superior, the man who can see the wood instead of just the trees, and gives it the reinforcement of the aloof detachment belonging to a court of appeal. The ordinary relationship prevailing between individual managers 'in charge of' different functions is one of rivalry, a rivalry which may be rendered innocuous to the persons involved by personal friendship or the norms of sociability, but which turns discussion about the situations which constitute the real problems of the concern—how to make products more cheaply, how to sell more, how to allocate resources, whether to curtail activity in one sector, whether to risk expansion in another, and so on—into an arena of conflicting interests.

The distinctive feature of the second, organic system is the pervasiveness of the working organization as an institution. In concrete terms, this makes itself felt in a preparedness to combine with others in serving the general aims of the concern. Proportionately to the rate and extent of change, the less can the omniscience appropriate to command organizations be ascribed to the head of the organization; for executives, and even operatives, in a changing firm it is always theirs to reason why. Furthermore, the less definition can be given to status, roles, and modes of communication, the more do the activities of each member of the organization become determined by the real tasks of the firm as he sees them than by instruction and routine. The individual's job ceases to be self-contained; the only way in which 'his' job can be done is by his participating continually with others in the solution of problems which are real to the firm, and put in a language of requirements and activities meaningful to them all. Such methods of working put much heavier demands on the individual. The ways in which these demands are met, or countered, will be enumerated and discussed in Part Three.

We have endeavoured to stress the appropriateness of each system to its own specific set of conditions. Equally, we desire to avoid the suggestion that either system is superior under all circumstances to the other. In particular, nothing in our experience justifies the assumption that mechanistic systems should be superseded by organic in conditions of stability.* The beginning of administrative wisdom is the awareness that there is no one optimum type of management system.

* A recent instance of this assumption is contained in H. A. Shepard's paper addressed to the Symposium on the Direction of Research Establishments, 1956. 'There is much evidence to suggest that the optimal use of human resources in industrial organizations requires a different set of conditions, assumptions, and skills from those traditionally present in industry. Over the past twenty-five years, some new orientations have emerged from organizational experiments, observations and inventions. The new orientations depart radically from doctrines associated with "Scientific Management" and traditional bureaucratic patterns.

'The central emphases in this development are as follows:
1. Wide participation in decision-making, rather than centralized decision-making.
2. The face-to-face group, rather than the individual, as the basic unit of organization.
3. Mutual confidence, rather than authority, as the integrative force in organization.
4. The supervisor as the agent for maintaining intragroup and intergroup communication, rather than as the agent of higher authority.
5. Growth of members of the organization to greater responsibility, rather than external control of the member's performance or their tasks.'[84]

CHAPTER 7

Working Organization, Political System, and Status Structure within the Concern

The last two chapters have been devoted to the exposition of two systems of management. The array of the five different systems found operating in various concerns (see Chapter 5) presented the state of affairs at points in a continuum extending from extremely mechanistic to extremely organic systems. Chapter 6 placed these empirical descriptions in the context of the relevant writings of sociologists and concluded with the formal description of two ideal types of system. In practice most concerns show features of both, but approximate wholly or in part more closely to one or the other ideal type as they adjust to conditions either of stability or of change.

In this chapter, we proceed to the effect on this adjustment of the working organization to conditions of change of the self-interest of members of concerns, whether they are pursued individually or in combination with others. For, conceived as 'pure' working organizations, concerns might be considered capable of adjusting their system spontaneously to fit their external and technical circumstances. Indeed, this is what we expected to be able to observe in the firms which entered the Scottish Council's electronics scheme. This presumption was based on the clear and important differences observed in the management system of the two firms of the preliminary study, the one manufacturing textiles in a very stable set of technical and commercial conditions, the other heavily committed to research and development in electronics. As the Scottish firms in the electronics scheme moved from stable situations to these typical of the electronics industry, we fully expected that obvious and important changes would occur in the management.

No such change of systems was visible in any of the Scottish firms. Instead, strenuous efforts were made to maintain the mechanistic system, or to reimpose it more stringently than had previously been the case.

The kind of development we met with may be made clearer by a description of the situation prevailing in one of the Scottish firms. The concern—an old-established engineering firm—carried on a large business in heavy fabrications. A new department had been added to deal with the design, production, and sale of technically advanced and fairly intricate products of a kind utterly different from those to which the factory was accustomed. Most of the components were made in other production departments of the works, this particular department having to produce designs, control progress through the works, buy in stock and parts, and run a separate assembly shop. In the recent past, the department had been meeting fairly serious difficulties in sales and thereafter in production.

The quotation which follows is a verbatim transcript from the record of an interview with a manager (to whom we have given a fictitious name). Previously a design engineer, he had recently been made responsible for technical sales.

Mr. Blake: 'Well, as it is to my mind, and I may be quite wrong in this, the most important thing in any industrial undertaking is delegation of authority and delegation of responsibility to the different levels from whoever is resident at the top. In our case it would be the managing director. Now the authority must be in line with the responsibility that you expect the individual to take. Unless you have that and you have one man holding all the strings and under him a sound team of people who are specialists in the various fields and the various management functions which they are to exercise, then you get what is happening here—it means you get a virtual break. I can only speak for the technical side, I can't say anything on the accounting side or the works side, purely on the design and development and the production functions that we have. Now, unless you've got this top management level with the delegated responsibility and authority, I don't think you'll be able to get a healthy industrial community. You cannot effectively skip that level and come down to the next lower level and make that directly responsible to the top or have odd little offshoots who bypass the

normal scheme and see themselves as responsible to the top, or altern-
atively, have somebody on the intermediate level who is not prepared to take the responsibility he should. . . . What happens in our
case, the way I see it, is that this top management level is missing,
and you're jumping virtually from managing director to a number of
senior technicians in charge of sections either on design, development,
research, or production, who haven't got anybody co-ordinating
their efforts except the managing director, by his occasional appearances and periodic meetings. That, I think, is the main trouble our
particular company is suffering from. Now, to take one instance, we
do have meetings: they are fashionable now. Something goes wrong
and then it's decided that a weekly meeting should take place.'

Interviewer: 'What kind of things went wrong?'

Mr. Blake: 'Well, supposing that a delivery goes wrong, or a
design is put into production and is functionally not quite what it
ought to be, then there is the usual post-mortem and it usually
boils down to that everybody realizes there hasn't been enough
contact. People have gone too far their own way and got rather
wrapped up in their own particular small aspect of the overall project. It's then realized that there hasn't been any one body drawing
together all the threads of the particular project and keeping an
overall eye on things, and then the usual scheme is to inaugurate a
weekly meeting at which various points can be raised and jobs can
be discussed. Now, in theory, that's a very desirable thing as long
as it discusses matters which are relative to its function. Also that
the plan of attack for the following week's work is clearly laid down
and it is stated who will do what and they report back at the following
meeting. It shouldn't just amount to a weekly gathering of bodies
who, halfway through the business, drink a cup of tea and then
break up and that's that. I'm afraid that is what has happened in our
case very often: it hasn't helped us much.'

Interviewer: 'What kind of things go wrong that are not put right
at such meetings?'

Mr. Blake: 'Well, when you start a new design or even a new
series of an established model there are certain steps you must go
through to get a new production batch on the go. These steps
involve the drawing office, purchasing, and, to a great extent,
planning and progress. That in turn will require examination of the
capacity available within the plant, machine loading and the rest,

and a great deal of ground work. . . . To get the whole thing going needs one instruction to one man who, in turn, will see that the wheels are set in motion. Well, that does not normally happen here.'

Interviewer: 'There is no one man?'

Mr. Blake: 'There is no one man. What it amounts to is that it's announced, "Wouldn't it be a good idea if we produced another ten of such and such a type?" and everybody says, "Yes, sir, it's a jolly good idea. Let's do that." The meeting breaks up and a few days later, for instance, I'm in the drawing office and happen to say, "What have you done about these ten units? Have you got the drawings printed? Is somebody getting the schedules out?" and they'll say, "Haven't had any instructions." I say, "Well, you go ahead and get the drawings printed and the schedules out." The chap in charge in the drawing office will take my word for it— *not* because at some time he had been instructed that I will give him instructions about what he had to do but because I've always done so off my own bat and I know that unless somebody tells him he'll stay put. That may be the first movement. The lot will go forward to purchasing. Stock control first will run through the schedules, tick off what is in stock and so forth.'

Interviewer: 'This is all routine passing of paper?'

Mr. Blake: 'There is an automatic passage of the schedule through the various departments. Ultimately somebody, usually it's the stores, will say, "Well, there's a hell of a lot of stuff in for this job. When are you going to start it?" Then everybody will think "When are we going to start it?" Because, at the time I said to Charlie in the drawing office, "It's time you got your schedules out, boy", nobody said to the planning office, "How are we going to fit this job in?" . . . People here are not pushing their jobs through but trying to drag them out at the other end, like a surgeon with a pair of forceps trying to drag the baby out. There should be no need to drag it out. It should come through.'

'Now it's true that in an organization of our size, with a variety of products and frequent change of machining requirements, you cannot plan as efficiently as for a company which turns out, say, electric motors or razors and nothing else. At the same time, each batch *could* be routed so as to ensure a steady flow. Secondly, there is the lack of co-operation due not to the individuals but due to the absence of functional contact. Production is based mainly on demand

—on contracts, not forecast sales—which means that when an article is wanted it's wanted yesterday. Then priorities are introduced or revised and any existing programme upset. . . .

'And our own departmental production manager has no direct control over machining facilities. Whatever he gets done in the shops in the factory he has to try to ease in and bolster his way around to try to get them in because there is nothing laid down. I mean, he would go and see the Works Manager and he could see the planning people, but the situation changes so fast; he wouldn't be in the picture all the time. They wouldn't keep him informed; they'd say, "All right, we can put so-and-so in the machine shop on; you can have them." And twenty minutes later maybe something else comes up. They take his job off and put the new thing on, and nobody would tell him. So two days later he'd come back and ask how the job was getting on. They might be sorry, they might not be, but it wasn't their affair really.

'Thirdly, lack of forecasting and market survey has also resulted at certain periods in our holding very large stocks and not processing them; in other words, capital being tied up in virtually useless material or part-finished products, or else in finished products lying on the shop floor in inadequate storage and of course deteriorating, which meant duplication of some of the work done on them—re-testing, repainting—before they actually went out.'

Mr. Blake, as we have called this informant, was feeling disgruntled, and this shows in the tone of his account; yet he had, just as clearly, tried to make the system work, and his account, moreover, squared with those of other members of the management staff. The problem is to account for the difference between the ineffectiveness of the organization here and the effectiveness of the arrangements in, say, the switchgear firm described in Chapter 5 (pp. 85-9).

Mr. Blake referred to a tentative move towards some kind of organic system after things had been going wrong. It had in fact been apparent that 'people had been going their own way and got rather wrapped up in their own particular aspect of the over-all project'. The device of a weekly 'progress and action' meeting had then been tried but, according to our informant, without much success. His explanation of the failure, it will be remembered, derived from his belief in the method of issuing commands and seeing they were carried out: 'To get the whole thing going needs one instruction to one man, who, in turn, will see that the wheels are set in motion.'

In essentials, the enlarged management meetings held in the switch-gear firm were a response to similar circumstances: things were going wrong because individuals had become preoccupied with their own particular roles and out of touch with each other and with the general task of the firm. And although the meetings were larger, the device of promoting an adequate flow of information and of obtaining agreement on future action through a series of meetings was, in common-sense terms, much the same.

But here the similarity ends. One indication of the difference shows in the kind of action taken *between* meetings and in the organizational context of such action. Mr. Blake suggested that unless explicit instructions were given by the head of the firm, nothing was done, or, if it were done, there would be a strong connotation of its being irregular, possibly damaging to one's position, certainly regarded as super-erogatory. In the other firm, difficulties revealed at a meeting were often met afterwards by arrangement between the individuals directly concerned and were expected so to be met. It is all too clear, also that the progress meetings mentioned by Mr. Blake were conceived by their members to be merely a means of assembling persons in the same room so that the mechanistic organization would have a favourable set of conditions in which to operate. The meeting followed the pattern of the foremen's meeting rather than of the Development Committee described on pp. 87–8. 'What it amounts to is that it's announced "Wouldn't it be a good idea if we produced another ten of such and such a type?" and everybody said "Yes, sir, it's a jolly good idea. Let's do that".' The 'sir' in the latter sentence is clue enough. In the absence of clear instructions from 'sir' nothing was done, except occasionally by Mr. Blake, and, presumably, others who were some-times prompted to act on their own initiative; but it is abundantly clear that Mr. Blake thought of himself as adopting a very irregular procedure in getting the drawing office started on the work required.

So Mr. Blake's desire for a clear-cut definition of his functions and of the lines of responsibility and authority about—and especially above —him may be understandable enough, but it cannot, as we have seen, be accepted as a safe indication of the way to more effective organization. It should be regarded as a reflection of beliefs about management organization rather than of understanding of organizational needs. And the situations in which such beliefs are commonly expressed—when individuals are frustrated in their attempts to get their own work successfully completed, when they are worried by the successful rivalry of others, when they feel insecure or under attack—these situations provoke an urge for the clarity, the no-nonsense atmosphere, of a mechanistic organization. It promises so many other dividends too. It is not only quicker to divide tasks into parcels, label them 'responsibilities', and post them to subordinates or other parts of the structure; this kind of procedure has the connotations of visibly controlling others, and the appearance of knowing one's own mind, which are valued aspects of executive authority. Conversely, one has the security of unquestioned power through orders to subordinates, the security of knowing the limits of one's responsibility and of the demands and orders of superiors, which the existence of something like Queen's Regulations can give.

There are other powerful motives for attempts to revert to a mechanistic form of organization, to return to a familiar system. Necessarily, when the market or technical base of a concern is changing rapidly, the situation is alive with opportunities for advancement and transfer, alive also with actual or potential threats to the status, power, chances of success or actual livelihood of some of the members of it. In Mr. Blake's firm this was so marked a feature of the organization that one junior manager confessed himself, along with others at his level, reluctant to seek promotion into a position in which he would be more dangerously exposed; he pointed for justification to the very high turnover that had been occurring among the senior management.

Under such threats there is bound to be latent or overt conflict between individuals or groups; and, as anthropologists have pointed out, hostility against persons in authority and aggressive action against rivals is often sanctioned by an appeal to traditional or other familiar values and codes of behaviour. What obtains elsewhere, and what used to obtain in this particular setting, is invoked to decide what should obtain here and now. In the situations encountered in this study, such appeals were engendered out of the disturbances set up by a changing situation, and comparisons were always made with the order of things in a stable situation.

PIECEMEAL AND UNNOFFICIAL MOVEMENT FROM
MECHANISTIC TO ORGANIC SYSTEMS

The need to adapt the form or organization may not necessarily appear only in the clear and decisive changes of direction such as the Scottish firms were trying to accomplish. Some large English concerns which had encountered substantial changes in markets and techniques became involved in a succession of adaptive shifts from one system to another which, being addressed to the expediencies of any particular moment, often moved in contrary directions, sometimes conforming with the general movement towards an organic system, sometimes going suddenly into reverse. This contrast is revealed by two situations which arose within the same firm.

Three elements in the situation were of fundamental importance: the amount of business done by the firm had expanded very rapidly every year since the war, and the firm had grown correspondingly; sizeable changes in the technique available for the design, and in the standard of performance expected of the product, now occurred, and revolutionary changes were impending; the people in management, especially at the top, had almost all grown up with the firm.

The account given here is taken from the report given to the management of the firm and discussed with them.

Some years ago, the Managing Director thought it advisable to create the new post of Technical Manager as deputy to himself. The Chief Engineer of the time was appointed to this new post, and the manager of one of the three Sales and Design divisions made Chief Engineer in his stead. Although the new post was a largely precautionary device, some redistribution was implied of functions, responsibilities, and powers which had previously occupied two people and now occupied three. This implied redistribution was never made explicit. It was not only left to the Technical Manager and the Chief Engineer to sort out the distribution between themselves, their mutual relationships, and their relationship to the Managing Director; questions of who, among their immediate juniors, reported to whom, how frequently, and on which occasions, were also left to sort themselves out into various levels of determinacy and clarity.

This account was given by more than one informant as the clue to any understanding of the management system. The same characteristic was, in fact, apparent in another way. In interviews, informants were often asked to explain the senior management structure. In no two cases did there appear to be agreement on who constituted the stratum of management under the head trio. A final decision was proposed by one informant on the grounds of membership of the management committee, which included the Managing Director, Technical Manager, Works Manager, the Managers of the three Divisions, the Contracts Manager, and the Chief Accountant. On the other hand, apart from those accounts which omitted several from this list, two informants placed the Production Manager on the same level as the Works Manager. Thus at the most senior level of management the status, power and, to some extent, functions attached to each person had a penumbra of indeterminacy surrounding what was unmistakably appropriate to his position.

This had two effects (or bears two interpretations) of which the people concerned were acutely aware, and a third which they naturally overlooked.

First, the lack of complete specification made for insecurity; no one at the top was quite sure how he stood relative to others; of whether one person's ascendancy was permanent or temporary; whether something which needed doing should be done by himself, another, or someone else; how far the distribution of functions, power, and relative status—as it appeared to him at the moment—was equivalent to the picture others had, and how changes which might occur would work out. Secondly, in spite of the anxieties of their situation as things now stood, most people seemed nevertheless not to want to get their position and that of others too clear. There was a kind of unconscious holding back from what looked an obviously desirable objective. For, once the position of everybody became clear, then one would lose one's freedom of manoeuvre, or the chance of adjustments somehow being made which might better one's position. A clear and definite distribution and specification of rights and duties would have frozen the structure, and nobody really wanted it frozen. (This particular observation was agreed later in two discussions with senior managers at the occasion of reporting back.)

Despite its embarrassments and anxieties it had nevertheless been accepted, and was at the time of the study regarded as efficient. The third point, therefore, is that the effectiveness of the top management organization was almost certainly greatly increased by each person's having this area of uncertainty around his own and other people's situation. With expansion and technical innovation proceeding at their present rates in the concern, new managerial responsibilities were bound to be frequent and largely unpredictable. In such circumstances, to operate with a clearly designated array of functions and responsibilities meant that any new feature of the total task of management could not be tackled until the Managing Director had become aware of it and allocated it to some one person or created a new functional position; a rigid structure, indeed, in which everybody knew his job and his authority and responsibilities and—what is more important—was clear about what he did *not* have to bother with, was liable to run into recurrent crises. Moreover, and equally important, the insecurity attached to ill-defined functions and responsibilities and status, by increasing the emotional charge of anxiety attached to the holding of a position, increased also the feeling of commitment and dependency on others. By this means the detachment and depletion of concern usual when people are at, or closely approaching, the top of their occupational ladder, the tendency to develop stable commitments, to become a nine-to-fiver, was counteracted. All this happened at the cost of personal satisfactions and adjustment—the difference in the personal tension of people in the top management positions and those of the same age who had reached a settled position was fairly marked. Such a cost seems, in the present state of knowledge about the effective operation of working organizations, to be an inescapable element of successful adaptation to growth and change.

The contrast to this unpremeditated and comfortless—but reasonably successful—adaptive shift in the system is provided by the course followed in one of the three Sales Divisions of the same concern: L Division. Until a short time before the study was made, L Division had 'looked after its own production', i.e., the design group had a major part in the production processes for equipment made to their designs. This meant, apparently, that a production manager appointed by the manager of L Division had supervised the manufacture of L Division's products in the works, and had been regarded as a member of the Division rather than of the Factory; while, that is, he formed part of the production staff, he 'reported to' L Division's manager.

After a period of rumour about possible changes, three moves occurred in rapid succession. The production manager appointed by L Division was instructed to 'report to' the Works Manager, thus removing the direct lien of the Division on production. The Works

Manager then replaced him. Thirdly, the Works Manager appointed a second manager with a 'general responsibility for the organization of production' for L Division. He also 'reported directly' to the Works Manager. This dual control system was later revised, the second person being made explicitly subordinate to the first, but until this was done these moves proved extremely disruptive. Among other consequences, production dropped alarmingly.

As one informant put it: 'The most appalling things went on—even records were said to have been got rid of by production people. I mean, you'd got in two new bosses who couldn't agree, who both went round picking faults, and so everybody started covering up like mad. Secondly, all the old intrigues broke out anew. I mean, blokes who had been demoted said, "Ah, here's a chance for me to get back into favour" —all the old nonsense. Then, as well as all that, there had been all sorts of subsidiary lines of communication between, shall we say, the office side and the works. Well, these started getting into frightful tangles. Human relationships which had been built up over a period of years ceased to exist overnight, and there was a good deal of bad feeling about it—in spite of efforts to soothe people on both sides—because there was a feeling in the sales department at one time to shop the lot on the production side—to prove them to be absolute bloody fools. Because you must remember that they had built up new markets which were going quite well, had got a lot of key contracts where we had got in for the first time; now we were going to be months late with deliveries and we were going to destroy the work of years, as far as the sales department were concerned. So there was a good deal of bitterness.'

Assuming that the account is a reasonable statement of the repercussions created—and it came from a member of the senior management not directly affected by the events—the significance of the episode lies not in the rightness or wrongness of the decisions themselves, but in the intensity of the disturbance they created. They appeared to have been far more severe than those which resulted from the not dissimilar episode affecting the higher management structure and certainly more than the effects observable from parallel episodes in other concerns.

The contrast between these two episodes suggests that the working organization, at the top and in the middle reaches of management, had evolved its own organic procedures. The moves in top management itself acknowledged—or took advantage of—this latent system. At lower levels, however, while the working organization had been made to operate effectively through the many 'human relationships' (informal channels of communication and commitments) formed between individuals and sections, the mechanistic system was suddenly revived with the redefinitions of responsibilities and executive functions. It became clearly the system which counted as far as careers and influence went; hence, for example, the sudden burst of activity directed towards the realignment of political and other ties.

While it is possible that part of the shock to the working organization came from the implication that an apparent trend towards the strengthening and semi-autonomy of the three Divisions was being put into reverse, the main trouble lay in the sudden reimposition of the mechanistic system which the management regarded as proper. This squares with the openly expressed predilection of many persons in senior and middle management positions for 'clear, tidy' organization, in preference to the informal and 'private' arrangements which were continually growing up.

Managers, even of laboratory groups, referred to organization charts showing the distribution of functions and responsibilities in detail. There appeared to be some insistence on clear lines of status distinction also. Within laboratories there were four grades of laboratory engineer: head of section, supervising laboratory engineer, group engineer, and laboratory engineer—over and above assistant Grade 1 and Grade 2. A senior manager referred to a 'rule' of having no more than four people reporting to their superiors. A production manager gave an excellent exposition of the way in which the principles of 'Scientific Management' should be applied in the factory, down to the methods engineer whose 'job is to do away with all skill of the individual'. Another spoke feelingly of the prevalence of the idea that 'a man's best friend is his telephone'. A manager, at least on the production side, should be able 'to cope with his own problems'.

These remarks point in the same direction as Mr. Blake's. Even though in both concerns there were signs of the spontaneous development of procedures typical of an organic system—some rudimentary, others fairly elaborate—there were recurrent endeavours to reconstitute a mechanistic system, to reassert some of its features, and widely-held beliefs that such a system was the only effective way in which to operate.

THE PROBLEM OF FAILURE TO ADAPT TO
DIFFERENT CONDITIONS

In the Scottish study we commonly found that, although the logic of the situation may have demanded that a firm alter its managerial procedures, no clearly formulated effort was made to do so. Indeed, it seemed to us that as often as not the head of a concern in a changing situation had successfully resisted all pressures to adapt the organization to the new situation; and in this he might have been wholeheartedly supported by the management staff, since they had responded to the increasing embarrassment of the concern and their personal apprehensions with more and more insistence on the need to reimpose the original order. Even attempts to introduce devices which might break down the authoritarian structure and serve as a medium for organic procedures might be thwarted by the single-minded devotion of a whole group of executives to clear-cut mechanistic procedures.

How is it that the working organization of a concern does not naturally adapt to changing circumstances? The answer commonly given to us by the people involved—the defective intelligence of colleagues or subordinates or the innate conservatism of the managing director and the 'old guard' of management—does not carry us very far. In every case that we know of in which this question can be asked it is possible to point to examples of technical and other competence displayed by the people most concerned, and to a positive interest in change which had at least carried the firm into the field of technical innovation.

Such answers assume that no other considerations apply than the welfare of the concern as a whole and the efficiency of the working organization. Yet manifestly, other considerations do apply. In this connexion, we may consider some general aspects of relationships between the laboratory group and the rest of the concern.

In all these studies, the most prominent aspect of this relationship seemed to lie in the differences between the ideas of industrial scientists, or development engineers, and the rest of management about their work and about how it should be done. These differences are discussed in later sections, but we may, perhaps, remark here that our first attempts at finding an explanation were towards presuming that they were symptomatic of two distinct cultures. The disparity between ideas of what was worth while in one's career or in one's life, or of proper conduct at work or elsewhere, of effective ways of going about a job, of acceptable demeanour, seemed to resemble the difference one expects to find between different nationalities, or between widely different classes in our society, or between milieux as different as those, say, of the merchant navy and newspaper journalism. We saw the management problem, that is, in terms of how to form a working alliance of a limited kind between existing management and a minority group despite their incompatibilities, or of how to absorb one into the culture of the other.

These 'cultural' differences were openly accepted by the senior managers with whom we were first in touch. 'Physicists', we were seriously told by one managing director, 'are very difficult people to work with'. But the same differences showed up in remarks to the effect that a good production engineer was a person who would tackle any problem given to him and solve it unaided, while a good design engineer in a laboratory was a person who could say 'I don't know'; again they appeared in references to 'long-haired types' and to 'the production clots'. The difference in the status system of the two sides appeared to substantiate this interpretation of a fundamental cultural difference. In the laboratory, virtually all members of design teams had a common professional status, entered by clearly defined channels of professional qualification by examination, a common educational background (in some firms, the usual term for a qualified member of the laboratory staff is 'a graduate') and often common interests and recreational pursuits learned in universities and technical colleges. Above all, they possessed in common a distinctive language and system of values by which achievement was measured. On the production side and in general management the men who had risen from the bench into management, the ex-officer trainees, the arts graduates, the accountants, production controllers, and technical and non-technical salesmen formed a far more heterogeneous system.

In attributing the differences between 'the laboratory' and 'the factory' to such a deep-rooted and conservative factor in behaviour as culture is normally held to be, we were mistaken. The mistake is worth recording because it arose directly from the statements made by laboratory people and production people about each other; the dissociation implicit in these statements was not only extensive and systematic, but was framed in terms of 'the kind of person' typical of the factory side or the laboratory side, so that we were led to think of explanations of the differences which would take account of apparent disparities in character. What we had not grasped was the tendency for people, when faced with problems in human organization of an intractable nature, to find relief in attributing the difficulties to the wrong-headedness, stupidity, or delinquency of the others with whom they had to deal, or more mildly, to irreconcilable differences in attitude and codes of rational conduct.

THE PRICE OF ADAPTING THE WORKING ORGANIZATION— AND REFUSALS TO PAY IT

In the Scottish firms which were entering the electronics industry largely as an escape from the stringencies of their existing commercial situation, the very first moves of the heads of the concerns seem to have been influenced by their first encounters with the foreign world of the development laboratory. These made them apprehensive about the impact of the newcomers on their existing staff.

During the period when they were exploring the possibility of entering the Scottish Council's electronics scheme, and informing themselves of the kind of work, and the procedures, which would be involved, they had made visits to other firms engaged in research and development and to Government research establishments, and the brief acquaintance gained thereby with laboratories and their inhabitants roused some apprehensions about introducing them into their own buildings and their own staff. 'The place was like a civil service set-up (*sic*), with chaps just sitting about, chatting.' Nobody seemed to have any responsibility—any feeling that he had a definite job to get on with and finish. 'I've seen some of those long-haired types—they amble about in white coats, come and go when they think fit.'

Secondly, there was the uncomfortable circumstance of the high salaries commanded by electronics development engineers with a few years' experience. While heads of concerns had come to accept the need to offer a high rate of pay as a necessary consequence of heavy demand and short supply, they were afraid of the reactions of their existing staff when they got to know—as they assuredly would—the difference between their pay and that of newcomers formally at their own level.

Added to these apprehensions was the fact that the managements concerned were quite in the dark about the technical nature and possibilities of electronics development. They had come to see that 'there might be something in it' for them, the electronics scheme appeared to reduce many of the risks involved, and it might be more dangerous to stay out of such work. But they felt almost helpless, and altogether too dependent on outside advice, when it came to the first really important administrative step of selecting and appointing a suitable leader for their new development group.

The way out of these troubles was seen by the heads of firms to lie in separating the new group as far as possible, administratively and physically, from the existing establishment. One firm planned to build a laboratory on the other side of a large field, decided that this was still too close, and bought premises in another part of the town. Another firm not only separated its new development group physically, locating it some miles away, but as soon as it was established, formed it into a separate company. In another concern, employing well under a thousand people, the head of the laboratory group, after two years, had never seen inside most of the production shops. In still another, the only party wall in the whole plant was between the research department and the production shops.

The primary object of the Scottish Council's electronics scheme was the exploitation of electronic techniques, learned by a firm through development contracts for defence ministries, in its particular sector of industry and in the markets which the firm felt able to explore and develop. This purpose could best be served, presumably, by promoting the fullest interaction between all the existing sectors of the firm—production, sales, financial control—and the research and development group, so that the latter could appreciate the needs of customers and users generally, the potentialities and limitations of the production shops, and the bounds of costs and resources, and the former could get to know the capabilities of the new techniques. Yet the first moves in most firms were aimed quite directly at preventing the establishment of those contacts through which the information vitally necessary to the success of the new venture could pass.

Of the three circumstances which have been enumerated as leading to the isolation of the laboratory team, the second, relating to differences in pay, is immediately identifiable as a threatened dislocation of the system of rank and privileges in the concern. Rather than have this happen, managing directors risked distorting or inhibiting the working organization necessary for the success of the venture.

The first situation is of a similar nature, though it is less immediately obvious. Managing directors were worried about the apparent disregard of obligations commonly accepted by industrial employees and about the absence of specific duties. The laboratory staff, of senior or middle management status, did not supervise a production or service section; they did not even have to produce specific information of guidance for immediate application. They ignored, or flouted, the prevailing modes of conduct in the firm. More privileged than others of comparable status, as they would be in the eyes of the rest of the concern, they seemed to want to make less return than others in the way of obeying the rules their colleagues had always accepted as proper. Segregating the laboratory staff preserved the *status quo*, so far as privilege and demeanour was concerned, of the pre-existing working community, at the price of effective organization.

The last of the three circumstances—the inability of the head of the concern to recruit a development team himself—meant the loss of some of the autonomy in decisions which had hitherto belonged to him. At the very outset, he was embarking on decisions about matters of which he was largely ignorant, and dependent, to a far greater extent than ever before, not only on the information but on the judgement of other people outside the firm. For some time to come, also, he would necessarily be in much the same position as far as decisions about development policy were concerned. The head of the concern, therefore, was involved in some personal loss of authority which, unless he was conscious of the modifications to the way of running the firm which a change of this kind might require and had made up his mind to them, could be construed as a loss of status. In effect, therefore, this circumstance, equally with the two others, reveals the way in which the proper functioning of the working organization can be inhibited by pressures deriving from the status system of a concern. He retained authority by arranging for communication difficulties between the research and development group and the rest of the concern, difficulties which helped to ensure that neither side would be fully informed when it came to major decisions and, accordingly, that he would retain the mastery in decision-making consonant with the head of the mechanistic system which had formerly obtained. An obvious and acknowledged need for the fullest consultation and co-operation between two sections of an organization could be entirely thwarted by a more strongly felt need to keep existing status positions and rank undisturbed.

By the time we made our first acquaintance with the Scottish concerns, the segregation of the laboratory group had often become established as part of the 'firm's way of doing things'. The dislocation of the working organization which was entailed had become translated into a number of separate issues between individuals or into recurrent difficulties. The order which had thus become established was regarded throughout management (including the laboratory) as reasonable and acceptable, defects being ascribed to the inadequacies of other people in technical skill, fairness, intelligence, personality, or social adroitness.

The translation of organizational difficulties surrounding the person into charges which can be made against others, or a technical problem of relationships into an emotional one, is a characteristic human process familiar enough when it is worked out in terms of national or international politics or inter-class or inter-racial conflicts. This is not to say, of course, that the reverse is true, that all emotional difficulties are resolvable into technical problems. Nor does it permit the emotional charges engendered by the process to be written off as irrelevant or superficial. People are often unfair to subordinates and others, ignorant where they assume knowledge, prone to see dangers or hostility in situations where none threatens, or clumsy and insensitive, or selfish, or lazy, and the people with whom they deal will dislike them for it. Yet what we have described was a general process repeated in a number of different firms, with very different casts performing very similar sets of parts. The recurrent appearance of a number of similar difficulties which were construed by informants in terms of the personal characters of other people suggested very strongly that the cause lay in what was observably common—the situation and the organizational vicissitudes of the concern. The origins of the trouble, also, dated from the introduction of the development group into the concerns, when, we must assume, there was a general willingness on the part of top management and the newcomers, to make the new venture succeed, and at least a chance of gaining the support of the other members of the concern for an effort aimed at improving its chances of survival.

THE MANIFEST AND LATENT ORGANIZATION OF THE CONCERN

The reason for the intrusion of these problems into a simple process of adapting the management organization to new conditions seems, then, to lie in the fact that the working organization of a concern is not the whole of the concern; it is only one aspect of all that goes on, and the ability of the working organization to adapt successfully to the conditions it finds itself in is limited by exigencies facing the people in it, exigencies which are only indirectly concerned with the effective operation of the concern as a working organization.

During recent decades we have become familiar with the idea that the circumstances of people's lives outside their workplace may affect their activities within it. But quite apart from the effect which events, attitudes, and beliefs arising in other social milieux might have upon the individual, there are aspects of a person's life within a workplace which are separate from his presence there as a member of a working organization. And apart, again, from what can in general be called sociable (or unsociable) relationships formed by people who work in the same place, the members of the working community which constitutes the concern are necessarily and unavoidably occupied (*a*) with the distribution and regulation of power in the concern, and (*b*) with their status in that community. The first set of preoccupations relates to what we can term the political system and the second to the status structure prevailing within the concern.

THE POLITICAL SYSTEM

The political system of a business concern is the product of the various demands, either actually and potentially conflicting, which are made on the total resources of the concern. Such demands can be framed in terms of money—for profits, wages and salaries, reserves, allocations for capital expenditure. Demands may also be for resources themselves, such as numbers of workpeople, assistant staff, space, equipment.

The advancement of self-interest by direct means is conducted openly in but few instances, although these may be the most important arenas of political conflicts (between management and workers, for example, or less frequently and less publicly, between directors and shareholders). Covert moves prompted by similar motives are, however, far more frequent and, indeed, play a continuing part in the life of the working community. These include the more easily recognized forms of 'empire-building' and the extremely complex and subtle manoeuvres practised by groups of workers in order to advance their claims, or to resist the requirements put on them by management and supervisors, or merely to gain strength against the day when it might be needed.

More relevant in the present context, however, is political activity which has to do with the amount of say which individuals or groups have in the destiny of the firm as a whole. The matter of conflict includes the degree of control one may exercise over the firm's resources, the direction of the activities of other people, and patronage (promotion and the distribution of privileges and rewards).

Political issues arising from this kind of endeavour are of very great consequence to the individuals involved in them, and to the firm. Their characteristic feature is that political ends are pursued as being in the interests of efficiency, or as contributions to the firm's task and its future prosperity. As in national politics, both sides to such political conflicts would claim to speak in the interests of the community as a whole: this is, indeed, the only permitted mode of political expression in the managerial legislature.

We do not mean to represent the conduct of management in firms as a continuous melodrama of hypocrisy and intrigue. These do, of course, exist, but the real problem, here or elsewhere, is most often that to the parties themselves, their opinions and policies seem utterly sincere and disinterested and their manoeuvres aimed at serving what they see as the best interests of the firm. The fact that the interests of the firm may best be served, according to the Sales Director, by building up the sales force, or, according to the Works Manager, by putting money into plant or machinery, or according to the Research Director, by taking on more development projects is not, they feel, material, and to impute that it is so is itself discreditable and unfair.

Whichever kind of political activity we are concerned with, however, demands, claims, or policies are usually put forward by different sections of the concern which recognize a common interest—shareholders, workpeople, office staff, production departments, the board of directors, or *ad hoc* groups—or by individuals representing them, such as heads of departments, Trade Union officials. The demands are put forward by virtue of claims recognized in our society as valid on legal grounds or on economic or other grounds which are socially approved. The political structure of a concern is the balance of competing pressures from each group recognizing a common interest for a larger share of all or some benefits or resources than they have now or think they may have in the future. There is, of course, little overt organization in the political party sense—although the alignments and manoeuvres of management and workers in some firms bear a distinct resemblance to party politics—but few members of working organizations inside or outside industry are unfamiliar with the other forms, the pressure groups and the enthusiastic individual seeker after power with his caucus of dependants or supporters.

What we are seeking to establish, however, is not that a political system or a status structure exists within each firm, which will be accepted readily enough, but that they do not exist as isolated entities. Political and status considerations constantly influence the working organization, and influence it so as to reduce its effectiveness.

Strategies in the Political Conflict of Management and Workers

A preliminary illustration of the distortion of the working organization by interference from the political system can be found in the familiar and obvious example of a bonus payments system. A large machine shop in one firm was manned with craftsmen who were regarded, both inside and outside the concern, as displaying high technical competence. Although design was reasonably stable, most of the machine runs were short; but there was no question about the ability of operatives to carry out the technical job of producing components of the required shape and composition out of what they were given in the form of material, drawings, and planning instructions. The firm operated a system of payment for work done of the usual kind, a basic time rate plus extra payments scaled according to the speed at which it was done.

The working organization of the firm, regarded independently of the political considerations we have discussed, assumed that instructions given to a craftsman in the form of drawings, planning schedules, and verbal orders or requests issued by foremen would be carried out at a speed determined by the craftsman's skill, speed of working, and desire to increase earnings. This assumption—again in isolation from these other considerations—would have been supported, in all probability, by every member of the concern, worker, trade union representative, foreman, and manager.

The 'basic time rate' is notionally the time estimated by a rate-fixer necessary for a workman of average ability to complete a single part. As elsewhere, however, this notional time was, in practice, fixed well beyond the mean time likely to be achieved.

The degree of difference between the rate-fixer's 'normal' times and the actual average was the direct product of the balance of political power between groups of men drawing such bonus earnings and the management. Thus we were told: 'This used to be a time-and-a-half shop, but during the war people got used to fantastic times. They would do a job rated at fifty minutes in ten. Now the norm is half time, and everybody reckons that jobs should give a bonus of 120.' When they were in doubt, rate-fixers would write in times a little less than what they thought probable, since they were certain that if the time was shorter than the accepted norm of half the average time, it would be disputed and 'they can always put a bit on; but you can never cut a time'. There was a constant traffic of men from the machine-shop coming to dispute times, with the foreman involved, usually either as an arbiter or as advocate for his workmen. This was thought of as being quite normal: 'You never get an engineering shop without cribbing.' If, because the material was defective or a previous operation wrongly performed, a particular machining job could not be properly done, men 'of course' worked on it and passed the part on since they would otherwise be paid for that period at the basic time rate.

The whole system was explained quite frankly, with the air of its being a normal and reasonable, though tacit, arrangement between management and workers.

Bargaining over rate-fixed times, as over bonus earnings in general, is a commonplace feature of what we have termed the political system. When earnings are determined by the extent to which a man can exceed a speed of work fixed by 'norms' which are the subject of informal systematic bargaining, then there should logically be some tendency for him, in concert with his followers, to demonstrate that jobs constantly take longer time, on average, than is allowed for. Provided that these demonstrations can be concerted fairly widely among the group, they will be able to enjoy increased earnings through an increased normal time' merely by the exercise of the political power they have within the concern, a power deriving not only from economic but from moral and other socially approved sanctions. It follows that for individuals to maintain earnings markedly above their fellows by working at high speeds is regarded as morally wrong as well as disloyal, unless it is sanctioned by motives accepted by the group as putting higher moral claims on a man. For such conduct can be used by management to demonstrate the fairness, perhaps even the generosity, of the 'normal' times fixed. 'Rate-busting', therefore, is reasonably regarded as disloyal behaviour and condemned in any individual worker by his fellows. He is undermining their case in a perpetual bargaining process, a process in which, time and again, increased earnings for all may result from the lengthening of rated times or raising of wage rates.

Within the limits set by the social, economic, and political system of which any factory community in Britain forms a part, the acceptance of bargaining in this particular matter serves the function of keeping any particular concern in being. Over a period of time there may be minor alterations either in the total benefits or resources to be disposed of within the concern or in the strength of conflicting forces. Thus the management may be in a weaker position than those of other firms because of its comparative ineffectiveness in political situations, or because working conditions are worse, or because jobs are relatively insecure, or because its need for labour is greater than that of other firms; in such circumstances the existence of an area like bonus earnings and rate-fixed time in which 'norms' on which pay is based can be adjusted so as to make up for the relative disadvantages is of practical use. Sunday overtime is used in much the same way. Similarly, fluctuations may occur in the amount of work coming into the concern, or in its profitability—and thus in the amount of benefits available—for all groups with a claim on them. Rate-fixed times can then be made generally 'tighter' as they were before the war or allowed to become 'fantastically' generous, as they were during the war. This kind of fluctuation occurs in addition to trends of a smoother and more explicit kind, extending over the whole industry, in basic wage-rates. Nevertheless, although bargaining over rate-fixed times fulfils a necessary function in the preservation of the firm's existence it also distorts the working organization.

In the case we have cited, for example, there was a direct incentive for workers not to report bad materials or spoiled work, but to work on it as though it were up to standard, and when they were finished with it, hope either that it would elude inspection or that they could earn bonus payment on the grounds that the responsibility for the defects was not theirs. Thus, on occasion, the interest appropriate to the individual as a member of the political system of the factory was directly opposed to the contribution expected of him as a member of the working organization—a contribution, as we have suggested, which he would probably have been quite prepared to acknowledge as proper, regarded separately from other considerations.

THE STATUS STRUCTURE

The second of the two systems of commitments existing side by side with the working organization relates to status. The status system is the way in which the rights and privileges, together with duties and obligations, are distributed among the members of a concern. In many regards the status system reflects the political structure. Yet the two are distinct, apart from the fact that political action involves combination, whereas in advancing or protecting his status a man may act alone. The reason for our choice of an illustration of the working of the political system at the shop floor level was to make it clear that a political gain does not necessarily involve any change in status in combination or alone.

It will be appreciated that status in a firm not only reflects the political structure but also the social structure of society generally. It is now fairly widely accepted that the class structure of British society is undergoing considerable changes. The two changes which concern us are, first, that class distinctions are approximating even more closely to a system of rank based on occupational position (which is in turn increasingly allied to educational achievement) and second, that there are far fewer of the grosser social frictions between classes characteristic of the past. The reduction of inequalities of income has much to do with it. Items of personal and family equipment and other symbols of status are more uniformly distributed, and there are fewer manifest discrepancies when people of different classes mix; it is no easy matter to distinguish the social class of members of a Saturday morning shopping crowd in the centre of the city. But by far the greatest change has been brought about by the segregation of people of different social classes into large, homogeneous residential areas, particularly in the suburbs.* Physical distance has replaced social distance. More recently, segregation has been augmented by withdrawal into the home for gardening, TV., 'do-it-yourself' pursuits, and other interests focused on the family and home. In general, therefore, class distinctions are less obtrusive in everyday life.

But within the workplace segregation cannot be practised. The educational and occupational basis of the contemporary class structure is presented in unmistakable and significant terms. Marks of rank are insisted upon, and often deliberately elaborated and multiplied; we

* There still exist in Edinburgh blocks of houses with a row of upper middle class dwellings on one side and, behind them, a street of tenement slums; but this juxtaposition is exceptional. Nowadays, an upper middle class family looks for a house not only in a 'desirable' street, but in a 'desirable' neighbourhood, surrounded either by equally desirable dwellings or open countryside.

have found wide agreement among industrial managers with the suggestion that the floor space of offices, the appearance, comfort, and cost of furnishings, the size of desks, the number and colour of telephones, height of partitions, the appearance of one's name in the internal telephone list or on the door of one's office, are all matters which have become more and more closely associated with rank, more and more used as expressions of one's position in the hierarchy of status.* This state of affairs has been contrasted by one informant with what used to obtain when the only important distinctions were, to bowdlerize his words, the separate dining rooms and lavatories for managers and for operatives.

The contrast of the multiplicity of distinctions and the frequency of contact between people of different status within the workplace and what obtains outside is not accidental. The workplace exists as the place in which occupational success—now the basis of social advancement in general—is won or lost. At all levels the identification of one's own relative status, in terms of rank and prestige, the appreciation of changes of status going on around one, and the definition of differences of status become, at times, of paramount importance.

This introduces a further set of preoccupations. Any one person is both a member of a working organization and one individual among many to whom the rank they occupy and the prestige attaching to them are matters of deep concern. Threats to one's status, the means of defending it against threats, and opportunities to advance it sometimes take on overriding importance.

Tactics in the Advancement or Protection of Status

Plain careerism may often be at odds with the needs of an effective working organization, and there are other ways in which the demands of the status structure can, broadly speaking, hinder or deflect the straightforward operation of a management system. But quite apart from the occasions when people candidly, cynically, or shamefacedly admit to having an eye to the main chance, and intentionally make decisions rather adverse to the concern's interests and favourable to

---

* The effects of this particular social change are visible outside industry, of course, as well as inside it. The department responsible for supplying furniture and equipment in at least one University now delivers 'professor-type desks' to appropriate members of the staff.

their own when there is a choice between them, there are very many more ways in which the affairs of the working organization are subject to control from the status system.

The two systems are, as we have insisted, merely two manifestations of the same actual instances of conduct. The manifestation uppermost is usually performance as functionary, but there are occasions when the two systems switch perspective, like an ambiguous figure, and the working organization becomes manifestly subordinate to status considerations. Two or three instances of incidents in which this reversal of precedence could be seen may serve to identify the very large sector of conduct in workplaces which may be classified as guided primarily by preoccupation with status.

This is, perhaps, most clearly visible in avoidance tactics. Challenges or threats to one's status are most easily and commonly dealt with by keeping clear of the aggressor. In one concern, all but the most formal working relationships were broken off for several months between the Chief Inspector ('Watson') and the manager of the Assembly shop after a quarrel between them. The Assembly manager had been harassed during the course of several successive meetings by the high figures for test failures and faulty assembly regularly produced by the Chief Inspector. 'We were going through a bad period', he said, 'but I got to be scared of turning up at the Monday morning meeting with Watson sitting there with his latest returns, and the boss obviously wondering whether there were any more screws he could turn on us. The trouble was that the figures were weeks late and no help at all in getting at the causes; they were just evidence of the department's failure to do its job. So I tried getting hold of the figures, on the quiet, of course. I spent a weekend on them and produced them on Monday at the meeting, with breakdowns which were definitely useful to us in production because we were reasonably near the time the things had been assembled. Well, that set us off, and there was a really heated quarrel between us, with the boss leaning back and enjoying himself—egging us on, almost. Do you know, after that, Watson never said a word to me for three months or more?'

The Monday morning meeting was not only part of the machinery of the working organization at management level; it was also a rink in which people displayed their suitability for promotion in front of the managing director. For the manager who told us this to produce inspection figures of a more effective kind in a much shorter time was obviously an improvement. But in doing so, he aimed a serious blow at the standing and prestige of the head of the inspection department. The move towards greater effectiveness could, therefore, be regarded as underhand and prompted by personal enmity. In the event it was so regarded. What happens in such situations is that information required for the proper functioning of the organization is not passed from person to person in accordance with the needs issuing from the tasks to be performed, but is used—or is thought to be used—to demonstrate superior worth or status.

Information, in fact, may become an instrument for advancing, attacking, or defending status. A production meeting was called in one concern to discuss the layout of a new production line requiring special methods. The meeting began with the layouts which had been prepared by a planning engineer (in his evenings) being passed round. Questions were raised about them and discussion went on for some fifteen minutes. The chairman, the Works Manager, then announced that the layout would, in any case, have to be drastically altered, and the usable space curtailed. There was a brief silence. When was this decided? At a meeting of the top management on Tuesday (three days before). After a further pause the chairman went on to propose an alternative layout. Discussion began again about the production methods to be used.

Afterwards, the planning engineer who had done the preparatory work said, 'It's a bloody waste of time. After all that talk—hours of it —all we got out of it was that we needed five benches. Well, we know that. And then telling us that we had to make sure that the assembly times were worked out beforehand. That's all theoretical; everybody knows it never works out like that. The date-line' (by which the department had to be ready to start production) 'is a piece of nonsense, and I could have said so months ago. We'll never make it by the end of the month. He' (the works manager) 'keeps at you, asking questions and makes you feel—well! Look, you're interested in communications. Well, we knew we'd have to expand for this new project and someone would have to move out. Then I hear through some individual in the factory, quite casually, that they're going to build a wall right through the department, and then I hear from somebody else that another bit of space was going to be taken up. I hear this through rumours, you understand, nothing more. Then I have to make this layout, and he says we can't have it that way now. So you get it again and again. It's no good. You heard he asked me for information about the number of operations. Well, I had it with me; I had it on a sheet. But I wasn't going to show it to him. I knew it would only start the ball rolling again with questions about the times and the methods . . . .'

Distortion of the working organization in the last instance had been carried to fairly considerable lengths. The flow of information from higher management down to individuals whose technical job it would be to translate it into workable plans for the layout of a department had been reduced to an inconsiderable trickle, and, during a critical period immediately before the meeting, stopped altogether. This had had the effect of maintaining or reinforcing the higher rank of the works manager, because the great disparity between what he knew of all the factors involved and what his subordinates knew, endowed him for the time being with the omniscience which is the fundamental requirement for a mechanistic system to operate. And it was so operating in a situation of considerable change, which clearly demanded organic procedures.

In concerns where the working organization is being distorted by what people experience as the prior claims of the status structure, the effect is for the type of procedure appropriate to the situation to be displaced by the other, for longer or shorter periods. The distortion is by no means always one-way;* in our experience, working situations requiring mechanistic procedures have frequently been replaced by inappropriate organic procedures because of the manipulation of information by subordinates in dealing with their superiors in rank. Indeed, there are good reasons why distortions both ways should occur in the same concern. It will be apparent that in concerns distorted in the way we have been describing, the proper discharge of a technical job by someone at, say, middle management level, will be more than normally dependent on the complaisance with which his superior regards him. There may be a tendency, therefore, for subordinates who are seeking promotion, or who are seeking to perform their technical functions

* The first years of any entirely new enterprise are often characterized by organic procedures, especially when it is small and growing; every aspect of its survival and growth confronts it with novel situations. Later, piecemeal or wholesale conversion to a mechanistic system takes place as novelty is reduced and routine becomes feasible. This conversion, however, may be by deliberate reform or unconscious evolution, may be bold or reluctant, early or late. For an account of the problems of such conversion in a case of arrested organizational development, see C. M. Arensberg & D. Macgregor[85] (cited at length by G. C. Homans in Chapters XIV and XV, *The Human Group*[38]).

well, to try to develop amicable relationships with their superiors. If they succeed in this, or are suspected of succeeding, there will be a corresponding tendency for those lower down, or those who feel that they have missed due recognition, to regard seniority in the organization as *prima facie* evidence of technical unreliability. There is an observable tendency for them to seek opportunities to betray the relative incompetence of superiors. This again, could be done most effectively by withholding information necessary for them to discharge their technical functions.

In more than one concern, there was observable a self-regenerating pattern of behaviour employed by some skilled men on foremen, and by some foremen on production controllers and managers, which consisted in an ironic style of address. This enabled the speaker to withhold any information beyond that specifically asked for, imply either that it was information that should have been known already to the enquirer or that the question betrayed foolishness of some sort, and so positively discourage further occasions of enquiry. In the lower ranges of the management structure, where mechanistic procedures are more often appropriate, such conduct exerted a steady pressure against the specification of jobs and the issue of definite instructions. Supervisors who yielded to this pressure were relieved of the need to reinforce their higher status by the display of fuller information. They submerged differences of rank in the 'free exchange of opinion and information' atmosphere which is characteristic of organic procedure.

Instances have now been given of the effective working of mechanistic and organic procedures in the conditions of stability and change to which they are appropriate; of attempts to apply mechanistic procedures inappropriately to the needs and tasks of a changing situation and organic procedures to stable tasks; of the distortion of the working organization by the political system when his membership of each system faces the individual with conflicting aims; of the distortion of the working organization when the need to defend a status or gain a higher one overrides the requirements of the working organization. Finally, we have suggested that such distortions manifest themselves as malfunctions of the working organization either through the inappropriate imposition of mechanistic procedures in a situation requiring organic procedures, or vice versa. We have now to examine the broader aspects of organization and of the situation of the firms studied in the light of these considerations.

CHAPTER 8

The Laboratory and the Workshop

In describing a concern in action as an interpretive process it was intended to give prominence to the co-existence within the working community of a large variety of technical and specialist 'languages'; those of the physicist and mathematician, of the cost accountant, of the draughtsman, the assembly-room foreman, the salesman, the fitter; and equally to the way in which things and events may have a large variety of 'special meanings' for these different people.*

The core of all the studies on which this book is based is the description of what happens when new and unfamiliar tasks are put upon working organizations. In many ways, therefore, the focus of the studies was in the one-way traffic of designs from laboratories into production shops, and many of the mistakes, conflicts, constraints, checks, and obstructions from which this traffic suffered had characteristics in common which can best be called linguistic.

The naming of different phases, or aspects, of the total interpretive process is itself a source of difficulty. The woman who comes in every afternoon to earn a few pounds a week soldering components on to a metal rack is doing a very different job, involving different technical knowledge and accomplishments, from that done by the members of a design team who worked out the design of receiver equipment for a radar navigation system, of which the rack and components form a

---

* The passage of work through a manufacturing concern may be regarded as a linguistic return journey, complementary to the abstracting process by which common sense observation of the physical world is transformed into scientific information, a process which is reflected in the transformation of everyday speech into technical language and mathematical formulae. Manufacture involves a return to the comprehensibility of common usage; this is particularly apparent in industries using advanced scientific techniques, and is particularly difficult.

part. Yet in what she is doing she is realizing the intentions of the designer; the *ways* in which she realizes those intentions have been devised by a production engineer, whose intentions she is also realizing. In fact, the activities of both kinds of engineer, and of the designers, draughtsmen, planning engineers, methods engineers, and others who have preceded the assembly operative in the interpretive system are all directed towards guiding or programming her activities. Their jobs are not finished until she has finished hers. The same interdependent relationships occur between any two positions in the system. And at any moment it may be discovered that intentions are not being realized further along the process, or that some intentions are incompatible with others—indeed, may be incompatible with the intention of the whole project in performance, price, or specification. Such moments are continually occurring in the process of designing and making new, complicated, and technically advanced mechanisms.

Not only is there a recurrent need for people concerned with one phase of the interpretive process to know what is happening elsewhere, the total process is interdependent and cumulative, being directed eventually towards ensuring that what is designed will be manufactured correctly, and that what the whole system produces will be in accordance with what the user required. As one industrialist said at the third meeting,* the range of activities in the electronics industry should properly be regarded as a 'continuum spreading from research laboratories to assembly lines and test benches'. If one cuts this continuum at any one point and examines the ends, it is difficult to describe the contributions to the end product made by the people on either side of the cut in terms which distinguish them.

RESEARCH AND DEVELOPMENT

Nevertheless, individual tasks and the technical language appropriate to them do differ. Arbitrary as they admittedly are, lines are drawn across the continuous flow, and the average level of each resulting section of the process thereafter represents a separate kind of activity which requires a different kind of skill and training, and a different language.

Nowhere was the difficulty of drawing these lines and deciding on average levels more troublesome than in the upper reaches of the interpretive process. Symptomatic of this was the difficulty over nomenclature we encountered in every concern, which made it to begin with

* See footnote, p. 94.

extremely difficult to identify and compare similar parts of the working organization in different firms.

For the entrepreneur (and the politician) it is possible to speak of 'R. & D.'—research and development—as a single, undifferentiated activity, ignoring wide differences of meaning and application, because the combined term refers to the use and cost of technical information as an industrial resource, as a factor in production. While this viewpoint is of central importance, there is a good deal of confusion prevalent in industry about what kinds of activity are comprehended within 'research and development', enough, at any rate, to make it desirable to attempt some clarification of what seems to be involved so far as our experience takes us.

On the one hand, there are plenty of laboratories and 'research departments' maintained by industrial concerns which are not in the least concerned with the design of innovations. On the other hand, we have instances of firms which have taken on people with higher degrees in natural science and experience of industrial technology in order to obtain designs of innovations, but which have, nevertheless, remained uncommitted to development work; their technologists have been looked to for drawing-board designs which are, indeed, improvements upon existing models, either of the firm's or of its competitors, but which are expected to be capable of being manufactured as they stand without any interim stages.

No precise definition is possible. All that one can say is that the whole process of applying new scientific discoveries to the manufacture of specific articles goes through a number of phases. How these phases are distinguished from each other, and what their true relationship is to each other, are recurrent points of debate among professional scientists and technologists. There is some general agreement that there is something called 'pure research', which is roughly equated in most minds with what goes on in university departments of natural science. Research of this kind is held to be distinctive in that it is carried on without regard to any possible usefulness it might have in the near or even the distant future. This image has been somewhat disturbed by the eagerness with which Governments have subsidized 'pure' research activities in physics since August 1945, but it still serves to distinguish pure research from 'basic research' which, although requiring acquaintance with the latest advances in pure research, is itself directed towards obtaining knowledge which will, it is hoped, prove useful. Typically, the research establishments maintained by the Government itself are held to be engaged in basic research, although some large industrial concerns claim that part of the effort of their industrial research workers is of this nature. Thirdly, there is industrial research, which bears on the application of the information thrown up by basic—or pure—research to the construction of devices or materials which have an industrial (i.e., a profitable) use. It is this stage which is in practice almost indistinguishable from basic research; it was clearly part of the activities of T.R.E. which were discussed in Chapter 3. It is distinguishable only because it is essentially a work of intellectual entrepreneurship—of applying the 'capital' resource of technical information to the creation of a utility which will satisfy a demand; it may involve looking for an answer to a definite question—such as finding out which micro-organisms could produce effects resembling those of penicillin, on which work the antibiotics industry is based; or Edison's search for a lasting filament for the electric lamp; or it may be a realization that scientific information of a theoretical or limited practical relevance can serve an industrial or commercial purpose, like atomic energy and the application of computers to the control of machine-tools.

Fourthly, there is the work of development, of converting an idea, technically valid and potentially useful though it may be, into an article which a factory, with its limited capacities of power, mechanical equipment, technical skill and labour, can make in sufficient quantities and sufficiently cheaply to make the whole venture profitable.

Lastly, and most confusingly, there is design, nominally the work of draughtsmen, and sometimes ascribed to them. It refers in general to the task of translating sketches, notes, models (and sometimes even prototypes for which parts have been machined in the laboratory's 'model shop' (laboratory workshop) or the manufacturing workshops) into drawings and lists of parts which will thereafter act as the definite guide and reference for manufacturing an equipment in quantity. Design is mostly compounded with development, the drawing office being part of the 'laboratory side'. Older professional engineers refer to this compound function as 'engineering'.

Research

A few of the firms studied maintained Research Laboratories. They enjoyed a good deal of autonomy, emphasized by physical separation from the rest of the concern. Most of their work was, in fact, done under direct contract for defence ministries. This means that Research dealt with the user on its own, without the intervention of sales and without the implication of Development or Production. In one company, indeed, the head of Research insisted at some length that it was not a service station of the 'operating company'. The research programme was settled in concert with the company's technical director in a fairly informal way, but these consultations were 'a guide for a continuing, changing, developing research programme'.

Research laboratories were also distinctive in their internal organization, teams being associated with techniques rather than with projects or products. It was said that the main difference between development and research was 'in the way we think. We (Research) don't think in terms of a product, but of a department of science'. In the same research laboratory, internal organization was said to be 'loose'. This meant, apparently, that while there were differences in seniority, each individual tended to regard himself, and to want to be treated, as a researcher. This made any organizational hierarchy impracticable; there had to be direct and easy contact between the head of the laboratory and each individual.

In many ways a special ethos of scientific professionalism was cultivated in research laboratories. One 'essential' difference cited between research and development people was that the former always wanted to take a problem to a 'final answer', whereas the development engineer might be quite prepared to surrender his part and interest in a project and switch to another. Again, it was 'policy' in one research laboratory to encourage publication of papers as a form of production proper to a research establishment, and perhaps as prestige advertising of a kind familiar in universities. In another there was a 'policy' of trying to see that each technique group had its own technical study project, this making for continuity and allowing 'scope for original creative work'.

Development

So far as Development is concerned, we follow what we believe to be most common usage, comprehending what we have defined as development together with design work other than that of draughtsmen and production engineers.

In contrast to research, development-design laboratories were much more preoccupied with formal organizational structure. In all eight English firms, for example, some twenty-four physically distinct establishments (i.e. research, development, or production) were visited. In only four of these did anyone refer to an organization chart; all four were development laboratories. In one laboratory, the organization chart revealed five distinct ranks, apart from technicians, who had three grades. (The existence of so many levels was said, it is true, to be a 'problem'.)

Here, as elsewhere, it is the co-existence of so many different 'languages' which creates a need for organizing. Organization in a development laboratory has to do with two things only: (i) the relating of everything that happens to the production of designs which will meet precisely and economically the requirements of users, and (ii) the support of the individual engineer by a network of communication with others which ensures that the technical resources of the whole staff of the establishment are readily accessible to him, and which yet does not require or encourage him to dissipate too much time or effort merely in communication with others. As laboratories grow larger, and specialist groups multiply, there is a danger of some essential channels of communication becoming attenuated or severed merely because of the presence of so many channels of communication around the individual.

Preoccupation with how to organize therefore, is directly connected with size in the minds of the heads of laboratories; indeed, 'organization' seemed merely to stand for the attempt to cut up big development establishments into groups of manageable proportions—sixty was sometimes cited as 'the maximum number any laboratory chief can cope with'.

Growth in scale in a laboratory has little in common with growth in scale of the rest of the enterprise. No work is replicated. More people means more individual, virtually unique, contributions of expertise, each of which may in time be of value to the work of every other member of the staff. At the same time there is, for the industrialist, the compulsive logic of the economies which large-scale should effect; this induces firms to keep large establishments together so as to allow of certain services, specialist technical groups, and expensive equipment being shared.

The experience of working inside one large development establishment organized into project groups, sub-groups, technique groups, and service sections, was described in these terms by one development engineer:

'You need to use such a lot of personal energy in order to get anything at all done. Take a simple example. . . . You want to make a piece of equipment you're developing. All right, you make some sketches, or get some sketches done, and then you take them along. You can't take them along to the model shop foreman and say, "Here you are, George, get these done for me." It has to go through the central planning organization. You take them to the central planner and immediately his reaction is, "Look, I can't do those for three months. If you want them done in six weeks you can get them done in six weeks, but you will have to go to someone and get priority on it." The general upshot is that you spend one day, or perhaps two days, arguing with first one person, then another, and as a concession you would probably get somebody to agree that you could have them done in a fortnight. It wouldn't make any difference, it would still take six weeks, but you would have spent two days arguing about it. Well, you get your bits and pieces made up and then you go to test them. Now I want some test gear. Can't find any test gear. Go and see the lab. steward. "Ah yes, what do you want, 'scope, signal generator? I expect we can find that for you", and you spend a morning looking round various labs. and you find the 'scope. You go to dinner, you come back, and find somebody has taken the 'scope, but in the afternoon you find the signal generator. . . . It really makes me wonder how the firm operates economically. There is an awful lot of time wasted in doing things that are unnecessary. The test gear situation is largely generated by unwillingness to spend capital, I think. . . .'

'In a large organization, too, you find that a project gets divided up and sub-divided up to a very large extent. You will find one lab interested in making the pre-amplifier. Another lab is interested in making the output amplifier. Somebody else fits these two together into an amplifying unit. Then yet another group in another room will fit that amplifying unit into the final equipment, whatever that is. Of course, that sort of thing happens on the production floor, but on the design (development) level it does tend to make communication difficult.'

This is a one-sided view. Even in such situations, more than one laboratory chief has since pointed out, there is almost always some other work to engage the attention of the individual engineer while he is waiting for essential parts or test equipment. Yet the account does reveal the tendency, which has gone quite far in some concerns, to 'productionize' development-design organizations, i.e., to break down the total process into stages akin to those of manufacture.

Despite all this, there was a universal emphasis within laboratories on the need for the fullest interaction between their members. In most interviews with engineers, too, there was agreement with the handy definition of a good laboratory engineer as 'a chap who can say "I don't know".'

They had to have some social skills . . . 'He's got to be able to work with people. That means, I'm afraid, that he's the kind of person you get on with yourself—not necessarily the "good mixer", the man who's been in all the university societies. You must avoid the genius who thinks everybody else is a clot.' The valuable man was the person who could say 'I don't know'. 'In this type of work, a man can't be self-sufficient. . . . There's got to be a little flavour of salesmanship about him; after all, he's got to liaise with components people, he's got to live and mix with model shop people and get them to develop confidence in him.'

By contrast, the good production manager was 'the chap who can get things done when there's trouble . . . and take instructions, not know better how to do it himself.'

In general, the circumstances of development work and the proclivities of engineers tend to make for an extreme kind of organic system embracing the whole establishment. As against this, sheer size and the traditions of industrial management make for attempts to dissect the system into sections according to one of two broad principles: 'teams' may be created to develop a whole system, or 'project'; specialist technique groups may be formed each handling a part of a project. In either case, there are residuary 'specialists' or 'project' teams. Neither altogether successfully avoids the wastage of effort by one group's duplicating the work of another. There is also another and more serious consequence which was remarked upon in the third meeting. It was suggested that the peculiar and vital organizational problems of the industry concerned the disposition of development-design teams rather than of the general functions common to all undertakings. Here the problems of balance, of emphasis, and of making groups elastic in size were always and inevitably present. 'Great efforts would be expended in building up the design teams of a division to adequate strength, with the virtual certainty that after a few months there would not be enough work to keep all the engineers busy. We are constantly coming across separate laboratory groups within the same company, one grossly overloaded and despairing of completing jobs in time, another obviously short of work and searching for slightly esoteric technical problems on which to keep going.' This was a typical situation, even where it could be met simply by the direct transfer of engineers from one group to the other. Because design was the core function of the industry, it was of the first importance so to organize the division of design effort that it could be moved to where it was most needed at any time.

PRODUCTION

The manufacturing side of all concerns comprised the kind of resources of skill and machinery usual in lighter electrical engineering, and the organization of them follows fairly familiar lines. Here we are concerned with the organization in so far as it impinges directly on other parts of the concern, especially development.

Typically, any problems or difficulties or malfunctioning occurring in an industrial concern give rise to a pattern of disturbance or inefficiency in manufacturing operations. In a manufacturing concern, as in an organism, difficulties anywhere in the system do not usually manifest until they begin to hurt: i.e. until they affect the manufacturing activities. Symptoms of trouble, therefore, usually make themselves felt first in production. Higher management, and other departments, commonly effect a kind of self-confidence trick in locating the cause, as well as the symptoms, of trouble 'on the shop floor' or 'in production'. It is therefore to be expected that production will appear to be in greater difficulties and subject to more criticism than any other part of the organization, and this has proved to be the case in the firms which have been the subject of this study.

This is not to say that the difficulties discussed in many interviews are unreal. Some measure of the internal problems in one factory appeared from the forty per cent allowance for direct labour made for 'contingencies' which was added to the rate-fixed times when making up programmes. (Direct labour costs, expanded in this way, formed the basis for all costing.) This allowance was normally found to be necessary and was used. It was ascribed by the production managers themselves to difficulties and interruptions arising from pilot runs, to stopping and starting jobs because of design problems, errors, and modifications, and to new and unfamiliar materials.

One critical statement by a development engineer, fuller and more elaborate than most, was made along the following lines:

Misreading of designs & specifications—'They are mostly people who are started off in mechanical engineering. Well, the whole of this stuff is black magic. They have in the past done things which to them seemed simple, like seeing a neat and tidy bundle of wires tied up with a string with one trailing outside the bundle, and thought this must be a mistake and tied it into the bundle. This was the one lead that had to be kept widely separate from the others, and they found that the thing didn't work. The designer then laughs heartily, and says, "You silly so-and-so's, of course that doesn't work" . . . this being extremely obvious to a fourteen-year-old schoolboy.'

Inability to detect minor errors in design—'They found that people working on waveguides for millimetre waves started asking for impossible tolerances, but when they go outside those tolerances they find the thing doesn't work. So eventually the whole thing becomes magic again. Whatever stupid thing a designer may have done, they feel they daren't change it; he didn't do it that way because he was stupid, but for some very special reason which only he is capable of understanding.'

Inability to translate a prototype model into a manufacturing design —'They feel it's the designer's job to spoon-feed them. . . . They should be capable of saying to the designer "Why are you doing it that way? If only you made it a slightly different way it looks to me as if it would serve just as well, and it would be five times easier to produce." . . . But our people take the attitude that this is the designer's responsibility; if he's a good designer he should know the way round all these production difficulties. . . . What the production people do is to take a design after it's finished and tell you all the things that were wrong with it and made it very costly to test and produce. They are never good at helping the designer to avoid this at the beginning.'

Unwillingness or inability to devise economical methods for small quantity production—'Generally, if you take a trained so-called production engineer and put him in an industry where you are asking him to find the most effective way of making fifty, he isn't interested. He'll spend his life feeling that you are cheating him, and if only you would make fifty thousand he could do them cheaply. He hasn't been taught to think that making fifty economically is a worth-while job.'

And inability to deal with jobs requiring special or novel production methods—'For instance, that in order to drill Bakelite printed circuit boards you have to use a certain kind of tip on the drill and run it at a certain speed, otherwise you split the board. This is a particular thing that happened: it's found so in the model shop and then it's put into the factory and they buggered up a whole lot of boards because it never occurred to them to enquire of anybody whether there was anything special about them. The designers hadn't written on the drawings "It is necessary to do these things in a particular way", and so the factory expensively had to find out all over again what had already been found out in the model shop.'

In some firms, criticism was directed more towards the transfer area itself: the Drawing office or Planning (production engineering) department was incompetent, or, more kindly, hopelessly understaffed. This could happen to the extent of its becoming a recognized sink into which resentments of a general kind could be discharged—once they had been translated into criticism. In one factory it was said that there was a constant stream of complaints from foremen about erroneous planning sheets, or the absence of sheets, 'You'd think people with any self-respect would get rid of these piles of paper on their desks.' The machine shop foreman was 'browned off doing the job for them. Who wouldn't get fed-up when a chap comes up with a tool drawing and says "What do you want this for?" and the whole 50-off job is finished.' But it was also said that foremen would 'race the planning boys' to make the first models before schedules were ready. And again, tools ordered might prove to be not quite right, or a sheet-metal job planned the wrong way round. At the end of this recital, the answer to the question 'Why not take these matters up with the Works Manager as they arise?' was 'We're waiting to build up a good case and settle it ourselves!'

There appeared, to many people, to be an obvious and direct incompatibility between the organization and methods of manufacture as they have been developed in the engineering industry over the past century, and the requirements of many of the tasks production is called upon to do in the electronics industry. Production engineering and notions about production management are both conceived in terms of large-lot production and fractional operations. Both are excellently adapted to the operation of mechanistic systems of management in stable conditions. This system is still the universal basis, we found, of the ideology of production management. We have already mentioned the exposition by one departmental manager of the principles and application to the department of Taylorism, down to the job of the methods engineer, which was 'to do away with all skill of the individual'. Others referred to 'personal' rules of having no more than four (or five) people 'reporting to' their superior. One insisted that research and development were services for production in the same terms as methods engineering and plant maintenance. 'If you let research take the lead, you fall to the ground.'

Yet the evidence suggests that the continued dominance of traditional production engineering principles is another possible source of strain, in that technical developments are, at present, leading in the opposite direction. Technical advances tend towards closer electrical tolerances; this makes for tighter tailoring of assemblies of parts made to what have been adequate tolerances. Since, for example, there is an increasing need for being able to vary inductances at the same time as maintaining stability, units of smaller and smaller value have to be manufactured to make adjustments possible. In order to remedy faults appearing in production because of overlaps or gaps in tolerances, and so that adjustments called for can be made in the course of production, there is, from the design point of view, a need for more and more skill at the work bench or as close to it as possible.

One informant, indeed, did suggest that it was becoming necessary to get the individual operative to contribute more. He cited an instance of a plating shop operative dipping parts singly, according to his instructions, while a supervisor was watching, and then reverting to a contrivance he had made for dipping a dozen at a time when the supervisor left. It was necessary, he thought, to enlist the ingenuity of the operative as a resource, and this could only be done by involving him in the technical and managerial processes of the firm.

THE PASSAGE OF DESIGNS THROUGH FROM DEVELOPMENT TO PRODUCTION

The criticisms by development engineers of the skill and managerial methods of production are influenced by their very different relationships with laboratory model shops. These are manned by picked craftsmen and are often large and expensively equipped. Normally, individual engineers are allowed direct access to craftsmen, a general supervision being exercised by a foreman and, more remotely, by the head of the laboratory; even where this arrangement does not obtain, it is regarded as the proper one, and attempts are made to approximate to it. In this way, the manufacture of parts and assembly of equipment is the product of close collaboration between laboratory and workshop staff, with the latter often contributing a number of ideas to simplify or improve the mechanical engineering of a design. The effectiveness of this relationship, the absence of administrative formalities, and the direct control available to the development engineer contrasts very strongly with what happens 'when a design goes into production'.

Indeed, the difference between production shops and model shops—the very existence of model shops—points to the incompatibility between the milieux and procedures of development and production. This incompatibility was underlined by the one or two cases where intensive co-operation, on small scale production lots, between laboratory groups and production had led to an approximation to laboratory-model shop relationships. In one concern, although a contract was for a single electronic equipment, it had been put into production shops hitherto concerned solely with electro-mechanical work 'so that the laboratory group can see it being manufactured' and also 'to build up production experience in this kind of work'. Manufacturing designs were matured by co-operation between laboratory and workshops, with new types of assembly shuttling to and fro, evolving towards printed circuits and other elaborations. Since production was involved so closely in the progress of the design, the department was not only able to work out ways of handling the problems of adaptation and to see the implications for the personnel of the department, but given a considerable interest in overcoming technical and managerial problems. On the design side, this meant that development engineers were able to call on increasing skill. On either side there was no inclination to treat relationships with the other as a 'problem'.

The electronics development group in this case was responsible for equipment throughout production and in fact until installation was complete. Not, we were told, that the principal development engineer had 'direct control over so many production bods'; he had to work through a production manager whom he referred to as a 'third party'—a minor but significant indication that it was a model-shop relationship that was, probably unconsciously, being aimed at. At all events, some success had been achieved in 'getting them to work to verbal informa-ation and rough sketches, as opposed to drawings, although 'they still regard themselves as production people'.

This kind of solution to the problem of incompatibility, which is certainly successful in small-scale production, is the most widely preferred solution to the whole problem. It underlies the practice of appointing product engineers from the members of development teams to supervise production of a design, and other devices whereby development laboratories can exert technical control over production. There are, however, major disadvantages to the general application of the procedure to large-scale production, most of them arising from the resistance of existing production management, and from the complication of production shop programmes that ensues. Also, the major difficulty is left untouched. Product engineers still have to work through the existing production organization, methods, and ideas.

Yet there were good reasons why production should take the attitude that a design when it came to them should be complete in every detail. The industry was founded largely on Government work, and officially, a completed design following upon a development contract is one which could be manufactured in any engineering concern with the appropriate resources.

Thus it was perfectly possible for another development engineer to describe the relationship with production in these terms:

'When the thing is designed we make prototypes and then drawings. These are used as the basis for production orders. These are then handed over completely to production, once you've got Ministry approval. Of course, you've had user trials, and introduced modifications, but eventually you get a sealed design, with numbers required and delivery dates.'

In fact, of course, this complete hand-over never seems to take place. Another engineer in the same laboratory could say 'The factory people seemed to think they were going to be a genuine production factory and to some extent they still have the same idea.' A colleague of his, quite independently having echoed this remark, sold the pass by adding, 'And, of course, we want to change things all the time.'

INTERMEDIARIES AND INTERPRETERS

The usual response to difficulties of passing designs through—correctly read in most cases as a problem of interpretation—has most usually been met by the creation of special intermediaries, whose job it is to interpret. In many of the concerns, there are groups of highly trained people whose existence depends on the continuance of these difficulties.

In one concern, the number of specialists acting as connecting links had grown rather formidably. The liaison structure began with sales engineers from product divisions resident in development sections. The model shops had been placed under the administrative control of production, the growing need for precision mechanical work in them being met by the introduction of methods engineers (from production).

After development had produced a satisfactory prototype, a job passed to the stage of production drawings. At this point again, production was involved, this time operationally, through more methods engineers resident in the drawing office, who 'supplied' production engineering technique to the designers and draughtsmen who produced drawings to the instructions of the development engineers. A fourth element which acted on the drawing office was the Standardization Group, which endeavoured to limit variations in the specification of components and the design of parts. This group was also 'resident' in the Drawing Office. The draughtsmen's job, therefore, was to see that drawings were produced by the date required by the product division, in accordance with the design group's specification, within the limits set by the standardization group, and in conformity with the ideas of the methods engineers. Quite apart from the organizational complexity which this involved, by detaching so many constituent functions from the draughtsman's task, his sense of involvement and his interest in his job was weakened, and his responsibility for seeing that drawings and schedules really did provide all the information they should, and were up to specification, might be surrendered to the supernumerary controllers who surround him. He became reduced, in fact, to what the managing director said he was: 'Just a chap with a pencil.'

Fragmentation of the process of designing and preparing for production continued in the next stage. Drawings, each signed by a D.O. methods engineer, passed next into production engineering (planning), and production prototypes were then made in the pre-production shops. At this point discrepancies often appeared between the performance of equipment manufactured to drawings and laboratory prototypes, discrepancies which could arise merely from the cumulative electrical distortion of parts and components which are individually within specification, or from misreading of design requirements at any of the preceding stages, or because of necessary adjustments made by engineers or model shop craftsmen to formal specifications and designs which are not entered on drawings. The traffic generated by these contingencies was again carried by methods engineers located in pre-production, who acted as intermediaries and interpreters as well as problem-solvers. A similar organizational arrangement was being built up in the production shops proper, with project engineers 'taking care of troubles arising with test, drawing office, and development.'

Development, finally, had produced its own liaison specialist with whom to confront the liaison specialist from the works. 'If there's a design going through production for which we've been responsible, well, a project engineer represents what happens in the works as far as we're concerned. He'll come to us with a tale of woe, and gets on to one engineer in our group—the post designs engineer. He may be able to solve the thing on his own. If not, he goes to the particular engineer who was responsible for that bit of the design. The works,' he hardly needed to add, 'are a *terra incognita* to us.'

This, the most extreme example, is yet only one instance of a widespread response to the problems of sheer translation which arise between people in different phases of the total interpretive system. It is an attempt to solve the problems posed by conditions of change by other means than the adaptation of the management system to an organic form. The creation of special intermediaries and liaison groups seems to offer the possibility of retaining clear definition of function and of lines of command and responsibility.

Yet it was not the manifest aim of management to revert to a mechanistic system. The head of the concern in which this state of affairs existed introduced his outline of the organization by saying that he, like the chairman of the group of which his concern formed part, was strongly in favour of giving jobs as little specification as possible. Doubtless, when every new group was created, a perfectly sound reason had appeared to justify it, and to show why the shorter circuit would not work. While it is not possible to point to the reasons in specific instances, we are inclined to ascribe the tendency to two characteristics of managerial thinking. The first is to look for the solution of a problem, especially a problem of communication, in 'bringing somebody in' to deal with it. A new job, or possibly a whole new department, may then be created, which depends for its survival on the perpetuation of the difficulty. The second attitude probably derives from the traditions of production management, which cannot bring itself to believe that a development engineer is doing the job he is paid for unless he is at a bench doing something with his hands; a draughtsman isn't doing his job unless he is at his drawing-board, drawing, and so on. Higher management in many firms are also worried when they find people moving about the works, when individuals they want are not 'in their place'. They cannot, again, bring themselves to trust subordinates to be occupied with matters directly related to their jobs when they are not demonstrably and physically 'on the job'. Their response, therefore, when there is an admitted need for communication, is to tether functionaries to their posts and to appoint persons who will specialize in 'liaison'.

One of the consequences which follows upon the confinement of movement and interaction by the creation of specialist interpreters is the growth of isolationist sentiment. There was, in our last example, a noticeable tendency to speak in terms of functions as quite detached from the commercial purposes of the firm; jobs were spoken of as discharged 'under pressure' from other groups, or 'as a service' to them.

SOLVING THE LANGUAGE DIFFICULTY

In general, the fewer the links in the chain from development to production, the more, that is, development and production were forced to learn each other's language, the more effective, speedy, and trouble-free was the passage through of designs. This was revealed with particular force in the one sector of the industry which is, significantly, wholly devoted to mass production and in which problems of the relationship of design to production, when they appear at all, take their place among a variety of minor endemic worries. Domestic radio manufacture tends to have few stages and plenty of interaction between them. Negative evidence of the effectiveness of this structure was obtained in one firm which had recently—as a consequence of a heavy increase in production load—inserted one or two extra stages. This was presumably partly a 'productionizing' reaction, arising from the notion that the kind of breakdown of processes which works for manufacture would help to improve the productive efficiency of design. It also incorporated another device typically thrown up by management as a response to overloading or malfunctioning: the introduction of several handover points. This fixes 'responsibility' and appoints firm commencement and completion dates for definable phases of a production design. This arrangement also encourages a group to inspect the work of an earlier phase for flaws before taking over responsibility.

The new system, with its six separate sections instead of the former three, was not only more complicated; it invited horse-trading in responsibilities and actively discouraged co-operation. In fact, the new arrangement was treated as irrelevant to the business of getting sets designed and produced on time and up to standard. There was just as much traffic up and down the new six-stage system as there had been before; the system had merely been made rather more difficult to work. The essential consideration here is that it would have been ruinous for the firm had the six groups 'worked to rule' on the new arrangement. For example, tools have to be ordered three months prior to the start of production, and if design work has not reached this stage, then the planning engineers said they might have to 'take a chance' on ordering tools, reckoning the gain of a fortnight or so worth the risk of ordering the wrong tool. Risks of this kind were reduced by planning engineers 'making it their business to keep in touch with all stages of a design'. The new formal structure of multiple stages was being ignored in favour of an informal approximation to an organic system.

Outside radio design and manufacture, relatively effective relationships appeared to exist in two concerns. Both were very young companies, and may have been still dependent on the sense of common purpose and the loyalty and enthusiasm which almost any pioneering activity can command among a group of people, especially when it is successful. Nevertheless, the problem of mutual understanding and adequate translation was being treated in both cases in terms at once more direct and more sophisticated than elsewhere. This, it seemed, was partly because the time when 'things worked beautifully and everybody mucked in' was so recent that the appearance of problems for the first time induced the firm, in analysing the problems, to ask why the original system no longer worked. The factory manager of one of the concerns gave this account of the origins of the present system:

'I used to live next door to the development director.' The kind of thing that happened was that 'say at nine o'clock one evening he'd get something down on a piece of paper. At ten o'clock I'd be with the pattern maker. At eleven I'd be talking to the production engineer. Next morning I'd be talking the thing over with my production people. By twelve we'd be looking at production resources and seeing where it would go in the programme. Well, as we grew, a lot of people were in the place trying to use the same technique without having the ability to put it through. So you'd get somebody outside—usually the accountants—insisting on coming in and formalizing the system by getting it documented.' Eventually, two years earlier, the company had engaged consultants to devise procedures for controlling the passage of designs through production. This was now referred to as the 'paperization phase' and was almost universally remembered as disastrous. 'They introduced a terrific amount of delay into the system. I suppose consultants are incapable of thinking about production in terms of tens and twenties off. . . . . It resulted in design groups, which were not affected by all this, continuing to produce drawings as before. These would get involved in a flow of paper from Banda machines. They (the design groups) would be called on to produce more and more drawings of sub-assemblies, etc., so that you'd get the process of piece-part manufacture and sub-assembly starting and stopping four or five times over, the stuff going into stores in the interval. This was very frustrating. . . . But we're beginning to see our way through this. . . . '

Although during the 'paperization phase' there was an attempt to interpose a clear hand-over frontier between design and production, with no works order issuable before a design was completely drawn, the system later regained its elasticity. Production engineers were attached in the early design stages to design teams. On occasion, they might be responsible for the production of the saleable prototype in the pre-production shops; they were further responsible for the production of a first pilot batch. This section had been given a rather higher status than is usual, and was regarded as the repository of manufacturing 'know-how'. They worked directly under the factory manager, who spoke of them as 'opposite numbers' to design teams. Their relationship with pre-production shops appeared to follow that normally existing between laboratory and model shop. The pressure for close interpretative interaction existed also within production. Production engineers, pre-production workshop, and planners 'lived close together' so that pre-production skills and experience with a design could infect the planning activity.

CHAPTER 9

Industrial Scientists and Managers: Problems of Power and of Status

---

THE INDUSTRIAL SCIENTIST

It will be remembered that although we at first accepted the distinctions drawn by industrialists between the personalities and the 'culture' of inhabitants of laboratories and the inhabitants of workplaces and offices, we were subsequently inclined to discount them as explanations of the difficulties which were found to exist. This was partly because such difficulties were not by any means confined to relationships between laboratory and management. Nevertheless, industrial scientists and other laboratory workers in industry do appear to possess in common certain characteristics which colour their relationships with other members of staff, and affect their conduct.

The status of industrial scientists and their value rests on membership of a social system of communication—that is, of a number of people between whom flows information about discoveries in the field of electronics, or biochemistry, or synthetic fibres, or nuclear physics, or aeronautical design, or metallurgy, etc., from one side, information about manufacturing processes from another, and about needs for new processes and products from a third. They become members of this social system by virtue of a tested ability to speak and understand the languages used in each of the three milieux which supply these three sorts of information, and by virtue of their ability to translate information from one milieu into information understood and usable in the others. This translation process is realized in the work of designing; of setting down comprehensive instructions for making a new product, or for making an old product in a new way, which utilizes information drawn from scientific discovery, has regard to the capabilities and resources of production units available, and meets the requirements of users concerning performance, reliability, and cost.

No peculiarity attaches to the technologist's being an intermediary. The entrepreneur stands in much the same position in regard to capital, labour, and market; the manager himself, in so far as he is distinct from the entrepreneur, is engaged in translating capital, labour resources, and technical information into products. The industrial scientist's special position is due to his semi-detachment from the industrial concern and to his new importance in the whole economic and social system of modern society.

The industrial scientist is now a member of a distinctive group in the national population, rapidly growing in numbers (which seem nevertheless to remain perpetually short of national needs), accorded a very considerable share of public and political attention, and inevitably conscious of the importance and the power attached to the technological contribution. This awareness has its effect on the general demeanour of the industrial scientist, particularly in his dealings with industrial managements, through whom the clamant needs of society for his services are expressed and who are the media by which these highly valued services may be translated into effective action.

To this general influence on the position and demeanour of the industrial scientist, two other relevant factors can be added. More perhaps in Great Britain, where the initiative in the expansion of technology has lain outside industry, than elsewhere, the industrial scientist conceives himself as a member of a professional group the bounds of which extend over the institutions of higher education and government service as well as of business.

Secondly, the technological system is new, and therefore fairly unfamiliar. In the greater part of industry, which had little direct dealings with new technology and industrial scientists before the war, preconceived notions about the kind of activity and the kind of person typical of the profession are bound to exist, and these have not necessarily been brought into conformity with reality. The lay conception of the technologist often derives from the earlier image of the inventor working on his own to produce fully-fledged innovations with dramatic suddenness, an image still surviving lustily, and maintaining a surprisingly wide currency. Jewkes, Sawers, and Stillerman, in their book *The Sources of Invention*, provide a recent example of the durability of this image:

'Inventors are very much a type whether they work inside or outside an institution. No one who has studied their lives or who has had contact with them can seriously doubt this statement. The inventor is absorbed with his own ideas and disposed to magnify their importance and potentialities. He tends to be impatient with those who do not share in his consuming imagination and leaping optimism. He runs to grievances and feels them sharply. More often than not he would make a bad business man. And so through a wide range of qualities of steadily narrowing generality. But, although of course there are not a few exceptions, his crucial characteristic is that he is isolated; because he is engrossed with ideas that he believes to be new and therefore mark him out from other men, and because he must expect resistance. The world is against him, for it is normally against change, and he is against the world, for he is challenging the error or the inadequacy of existing ideas. It is precisely because of these eccentric qualities that society had always found it so difficult to fit the inventor into its scheme of things. Gifted so highly in many rare ways, he is often oddly devoid of worldly knowledge and thereby more than ordinarily in need of assistance. Yet that assistance must be given largely on his own terms and in a manner which does not frustrate or destroy his originating powers. He is capable of self-deception yet he can be right when most others are wrong.'[86]

Industrial scientists are not disinclined to accept this image, with the flattery it implies, and a reputation for eccentricity may be helpful not only in creating a complaisant attitude in industry towards the transgression of the normal rules and towards claims for privileges, but also in furthering a career. Certainly there are already a number of 'legendary figures' in industrial research whose tolerated and even cherished penchant for flouting conventions has added to their reputation. The 'prima donna' scientist is a much more familiar figure in the electronics industry than he appears to be in the academic world. Nevertheless, the prevalence of the image, and its acceptance by some industrial scientists as a model, is also taken to imply in some degree abdication from claims to leadership in the general management of business concerns. The identification of a person as a 'scientist with a touch of genius' or a 'university professor type' has become, in more than one of the firms we studied, an effective means of excluding him from a share of managerial control.

FRONTIER CONFLICT BETWEEN DEVELOPMENT AND
PRODUCTION

Development and production meet at the Drawing Office. It is at and around this point that the major linguistic difficulties occur. Draughtsmen vary from firm to firm, and even from job to job, in the amount of direct guidance they require from development engineers and from production engineers. There are times when production and planning engineers can save time by getting information about a design directly from laboratory engineers, rendering the draughtsman redundant, apart from the value of his drawings as records. More frequently, however, the unfamiliarity or novelty of laboratory designs defeats his capacity to translate them into true manufacturing designs. We have already described one attempt to meet this need by inserting extra specialist interpreting groups throughout this system, and especially around the draughtsmen (p. 169). In more than one instance, similar attempts at solving the translation problem in mechanistic terms have met with opposition from one side of the frontier or the other, or have themselves become arenas for political conflict.

The interplay between the hierarchic structure, with its constant thrust towards the resolution of organizational problems in mechanistic terms, and the requirements of changing conditions for organic procedures; the attempts by laboratory engineers to substantiate their claims to higher status within the firm; and the struggles by powerful groups to establish control over disputed areas of the concern can cumulatively render the situation at any one time extremely complicated. Conversely, people's own appreciation of the situation in which they are placed, and their own actions, may also be construed as elements of various constructions we have put on them—as features of the working organization, of the status structure, or of the political system. This interleaving of overt management issues and latent preoccupations with status and power, and of the problems posed by linguistic differences and the opportunities for manoeuvre which they provide, may be demonstrated by developments in one of the larger concerns.

It was generally accepted throughout the concern that it was divided into two parts: the 'design side' and the 'production side'. The 'design side' consisted of several teams made up of 'design engineers' and technicians, while the 'production side' included not only all manufacturing departments but also the drawing office and the machine shop and workshops serving the laboratory groups. According to a principle often repeated by the head of the concern and by senior managers, responsibility for the design of all products made in the factory lay with the laboratory chiefs, while the chief production engineer was responsible for manufacture. Since the whole plant was geared to the design and manufacture of electronic devices and systems of an advanced kind, the interpretive process covered an extensive range of 'languages'. Things expressible only in mathematical and engineering symbols comprehended by a very limited number of highly trained specialists became things made and put together by craftsmen and unskilled operatives. In the eyes of management, this meant that there had to come a time when 'design brains'—the people qualified and 'responsible' for producing an equipment which will serve the purpose proposed—handed over to 'manufacturing brains'— the people qualified and responsible for turning the equipment into a product which could be manufactured (and manufactured at the least possible cost). Yet in effecting this transition, the design might not be so altered as to affect performance. On the other hand, for the transition to be possible at all, there had to be some foreknowledge of the capacity of industry to manufacture the equipment within reasonable cost limits. So while there might have been in principle, a 'point' at which the design side handed over to the production side, there had to be interpenetration of the knowledge and skills of both to the beginning and end of the whole interpretative process. It was no use thinking of a project unless there was some possibility of manufacturing it. On the other hand, the design side had to be satisfied that products sent to users were as effective as they had meant them to be, and this, with complex products, meant not only designing test equipment but seeing that the exigencies of manufacturing and the lack of their kind of technical knowledge did not introduce changes which would affect performance.

As the management of the concern saw it, the need was for answers to two questions: at which point did a project become, in the main, a job for the production engineers? and what steps were needed to ensure that manufacturing as well as design requirements were taken fully into account when the design was drawn?

The practical problems which had existed in the past were described by the head of the drawing office in an interview:

'There was the big problem of having physicists and so on pretty well straight from the 'varsity and in many cases not having been in an engineering factory before, designing things on the drawing board with the aid of a draughtsman who thought he knew all the answers.'

'The draughtsman would work directly with . . . ?'

'The draughtsman at that time did work just directly with the physicists and some of the results were startling, to say the least of it.'

'What sort of things happened?'

'Well, one thing that happened was that the draughtsman produced things that were extremely difficult to make and in certain cases impossible to make, and wouldn't do the job; they worked in many cases to extremely tight tolerances, you see. The physicists would ask for a certain accuracy and the draughtsman would promise them pretty well anything, and when it came to making the job, it was found just utterly impossible to do this and we found that we didn't have sufficient tolerances to make individual parts. And the accuracies the machines could work to, and so on. . . . In these days when a job was developed in the lab it passed completely from the lab to production all at one go. We'd develop such a thing as a 10 mm. bi-polar whirligig—we'd develop and make a model in the lab workshop. This would be tested in the lab and found to be all right. Then we might make about half a dozen in pre-production and test them in the lab. They would all be hand-fitted, hand-made, and found to be fine. Then we'd take the job and decide to make a thousand off in production and of course the production engineer immediately rejected most of the drawings as being unfit—absolutely unfit—for production. The result was you got a tremendous gulf between the experimental side and the production side.'

This 'tremendous gulf' was seen, rightly, as a linguistic problem. The inability of the people previously concerned to translate directly from laboratory models and prototypes into manufacturing resources was solved by the creation of a group of interpreters: 'Now the obvious way to overcome this is to have a product engineer starting with the job at a very early stage, and he would advise the people in the lab on how it should be made to suit production, and he would come with the job right through drawing office, lab workshop, pre-production and so on right on to final production of the job, controlling it all the time.'

This 'Engineering Department' was located strategically in the drawing office itself, on the side by the door leading from the laboratories. It was also, of course, part of the production side. Both these facts carried implications. First, the use of the new group to control interaction between the laboratories and the drawing office and workshops curtailed the amount of control exercised by the laboratory groups. In particular, they lost the right to call directly on the laboratory workshop—their most essential factory service, since the contribution of the new engineering department was regarded as most necessary at this point, when prototypes were being made. Secondly, the inclusion of the new group within the 'production side' brought into prominence the political elements of the problem, in that the 'production responsibilities' of the new group were still clearly differentiated from the 'design responsibilities' which remained with the laboratories; the structure of command was unaffected, and the presence of a new force near the frontier, so to speak, stressed the line of demarcation.

These implications made themselves apparent in a number of ways in the course of interviews with laboratory staff. For instance, requests by a project team in a laboratory for drawings or for machine-shop or fitter's work had now to be passed through the member of the engineer group who was attached to the project. As one team leader said, 'Chaps used to wander back and forth between here and the lab workshop and complete a job perhaps mostly in the lab; there was a card to requisition workshop time which had to be signed by the lab chief. This sort of procedure is now dead. All work is supposed to go through the engineer; there has to be a drawing and so forth, and the whole process is getting rather elaborate. Of course, you get a well-finished job, but you get it weeks too late. It has to go, for example, through the drawing office, the rate-fixers, then get put up on the load-board—the whole administrative bag of tricks, and it takes much too long.' There was in reply a deliberate effort to build up resources of skilled mechanics and of machines in the laboratories, and so remove the need for going through the new procedure; some fitters from the workshop had been acquired 'on permanent loan'. Backing this development was the experience already gained by some members of design teams in affecting the transition from design to manufacture; some claimed to know as much about the relevant production methods as the new engineers. Furthermore, in one laboratory section, a

production engineer had already been appointed to deal with manufacturing problems arising during development work; he was called
by a team leader in another section 'an engineering group on his own'
—they were considering the creation of a similar position in their own
section. There was therefore no need to refer all problems relating to
manufacture to the new engineering group. Complementary to this
pressure towards making the services of the new group partially redundant was the practice of using them as providers of specialist contributions; the engineer was usually eager to prove his professional status
on a specific job, and this kept him out of harm's way on the periphery
of the design team's work.

More fundamentally, senior members of the laboratory staff maintained that it was not possible to build up separately on the production
side knowledge and experience of manufacturing technique which
could then be 'fed in' to a design, or for this process to be performed
so that the product engineer could then control manufacture. As
one of them put it: 'Production—proper production—is only possible
when somebody from the design team goes through with it. This is
what's been happening in this factory for years—I'm talking about what
happens, and not the theory of the thing. You can't build up from the
other end. They're trying to produce engineers to do it, but I just
don't see it happening. For instance, the engineers have never been
inside this lab.' He went on to say that the only cases of people 'going
through' with a project had been when members of the original team,
some of them not up to the standard of technical ability prevailing in
the laboratory, had transferred from the design side to the production
side.

This last informant's remarks disclose a latent preoccupation with
issues other than those of effective working. He first claimed that the
job of 'going through with a design' could only be done properly by
design engineers themselves, and then deprecated the standard of
technical ability appropriate to the job. The implicit purpose of the
statement appears to have been to assert the higher standards of all-
round technical ability prevailing in the laboratory, a superior capacity
which entitled its members to control the making as well as the design
of products.

As the context of the new engineering group was explored, so it
became more and more apparent that the problems of the interpretive
system were parts of a larger complex of problems of status and of

politics. The laboratory staff, from their own account, were attempting to reduce the effectiveness of the new arrangements for improving the transfer of essential information from themselves to the production side. The arrangements themselves were shaped to some extent to fit the main political divisions of the factory, and to play their part in the conflict between them. The question arises, indeed, how far the defects in communication which had appeared to call for the institution of the new engineering group found their own origin in the social and political division of the factory into two 'sides' rather than in an insufficient supply of the right kind of technical knowledge at the point where lab and workshop met.

ISOLATION

The simplest and most direct consequence of the special status ascribed to the industrial scientist is that in extreme cases he becomes so detached from the rest of the concern that it derives little benefit from his presence. This was the situation in one of the Scottish firms. It had been devoting a good deal of energy and money to the exploration and development of new markets. A year or two before we became acquainted with it, an industrial scientist was brought in to take charge of a 'Research Laboratory'. In addition the firm had the services of academic consultants, and was supporting research in a University department of Engineering.

'The main idea in starting the Research Department', said the industrial scientist, 'was that it should pick up on development work which other firms in the industry seemed to be pressing on with.' This development work included electronic control devices in which notable advances had been made by one or two major competitors, but the projected scope was much wider.

At the end of two years, virtually no development work had been initiated in the laboratory. The small 'R. & D.' team had become a test laboratory and a group of consultants on production problems. The main feature of the laboratory was the test gear, which the firm had decided to install after the appointment of the research manager, who had been responsible for a similar, though much larger, plant in his previous job. 'There was no intention of getting this plant in when I applied for the job.' He was involved in testing throughout the production process, and spent a good deal of time on incidental problems of the design and manufacture of items of current production referred to his department. When we suggested to senior management, after some time in the works, that a tendency seemed to have grown up for the factory's design and production departments to unload on to the research group problems they would formerly have tackled themselves, they agreed that this was so.

There seem to have been two main reasons why the development laboratory found its work diverted from what it had been created to do, and its growth stultified.

(*i*) The first was the existence of arrears of development problems arising from users of their products. The industrial markets now being served by the firm were continually throwing up needs for improved types of equipment. 'Over the years the firm's collected a number of jobs we've quoted for,' said one senior manager. 'The policy has been that whenever the Managing Director has had a letter from an outside customer enquiring about something not in our range, and he thought we might do something with it, he'd insist on quoting for it. The same with tenders. Well, we'd do this "with a pen and a piece of paper" and send something in. Usually we never got the order. If we did then everybody would be running around in a panic getting drawings done, trying things out and so on. The upshot was that we acquired a long list of things which at one time or other we'd decided we might try.' Before the advent of the Research Department user problems of this kind and others arising more directly out of the firm's experience with customers' needs had been a general management task; in particular, the Managing Director's considerable flair for production design had kept the list of outstanding items turning over, albeit more slowly than the rate of arrival of new problems. After the Research Department was set up, the Managing Director maintained his practice of attacking this list of items at places where he could see some possibility of a profitable solution, but without much regard to what the Research Department was doing.

'When they (the "Research Department") moved in, the Managing Director was very eager to come in and do a lot of work. "Now," he said, "I can really get down to the things I'm interested in," and they reserved a small office for him upstairs and he came in and supervised the finishing of it, got all the pencils sharpened on his desk and so on. He's never been back there since.' The number of items on the list, their specific nature, the firm's tradition of 'trying its hand' at problems posed by actual needs arising in the market, and the tendency to use the Research Department as a technical reinforcement, all combined to direct the main exploratory effort of the laboratory in the same direction. 'There's no policy about development work as against this dealing with items and small enquiries.' So the position was that there were two fairly distinct attacks being made on the list of development ideas derived from market contacts. They were conducted at different technical levels, and the angle of approach was very different; the oversight exercised by a tightly organized senior management group was quite sufficient to avert waste of effort through repetition. But the creation of the Research Department did not in fact introduce any innovation in the firm's products or manufacturing methods of the kind explicitly looked for when the Research Department was planned. Instead, the added resources were used to reinforce the previous efforts to improve design and meet customers' special needs, and to supply better test facilities. Moreover, when the next large expansions of the firm took place, they were away from innovations: the firm bought control of two other concerns, one producing castings of the kind used by the parent firm, the other producing components. In neither case were these concerns converted into early process departments serving the parent firm exclusively. It was planned to develop each within its existing market, but it is significant that, despite the knowledge the firm had of the pace of technical development in the industry, the first large-scale investment of capital and managerial effort for years was put into expansion of production resources, and into selling the end products of the concerns acquired.

At one level of explanation, then, the sheer momentum of the firm's previous practice in building up and working on a development programme may be said to account for the distortion of the working organization needed in order to operate the new plan for development work. Yet the firm had a bent for expansion, for taking on new functions, attacking new markets. The working organization proved itself, time and time again, extremely flexible and readily adaptive to new or recurrent needs arising from sectors other than research and development work; indeed, it is this firm which has provided us with some clear and important illustrations of the working of adaptive organic procedures in organization. The question remains, therefore, why the distortion took place, but now takes the form of why it was that the organization of research and development *in particular* remained stable despite a conscious effort to change it.

(*ii*) A number of circumstances seem to add up to sufficient evidence for taking it that the second and fundamental reason lay in the peculiarities of the status system.

A great deal of emphasis was placed by senior management on the possession by all members of the sales, engineering, and production staff, of a common language, common background, and common level of technical competence; almost all had qualified in the same technical college. The management staff, therefore, although at different status levels in the concern, nevertheless possessed a common professional status. This was somewhat lower than that of the Research Manager.

The Research Manager, after two and a half years, was the most isolated member of the top management group (of thirteen members). This appears from a study[87] made of the actual expenditure of time by each member of this group according to seventeen subjects and according to the persons with whom they were in contact. The Research Manager was engaged in the smallest range of subjects, apart from one senior office manager. He communicated with the smallest number of other departments. He spent the greatest amount of time of all managers in his own office and department, and the least amount of time with other managers at his own level or above. When each participant in this study was asked to ascribe a rank relative to his own to each of the other participants, most difficulty and most difference of opinion was found in fixing the position of the Research Manager, who was finally given, by agreement, a position of his own below that of the directors but above the seven heads of departments who were of similar age.

The research department alone was physically separated from the rest of the works by a wall.

The reason why the inertia of previous practice in the firm seemed to apply to the research and development programme and not elsewhere appears to us to lie in the circumstances just outlined. They point to a rupture between the Research Manager and the rest of a tightly integrated management. This rupture is closely identified with the distinctive status which attached to his possession of rather higher technical qualifications and standing than those common to the rest of management.

AVOIDANCE

In the concern which figured in the previous section, there was an actual brick wall between the laboratory and the rest of the concern, the only barrier in the plant. In other places, the 'brick wall' which was mentioned in interviews and consultations was metaphorical, but even so was at least as much of a barrier.

In a large research and development organization, employing scores or hundreds of industrial scientists, there is ample room and opportunity for the cultivation of the development engineer's self-conception and of the claims to special status which arise from it. Conscious of the social and economic significance now ascribed to his qualifications, linked by training and occupational interest with colleagues outside industry, credited too, perhaps, with some of the supernormal abilities and abnormal mannerisms which have figured in popular myths about the scientific inventor, the industrial scientist is under some pressure to regard himself as apart from the run of salaried employees in industry.

If he subscribes to this view of his apartness, then the industrial scientist will incorporate in his conduct a number of signs which will separately and in combination demonstrate this to other participants in the occasions of his working life. Starting work an hour or so later than the production and office staff is one such sign; it derives not from assertion of an occupational inability to get up early but from the general specification of scientific work as erratic and subject to varying rhythms of activity and to bouts of enthusiasm which, by compensation, *may* keep a worker in his laboratory into the small hours or over the week-end. The cultivation of eccentricities in dress and in leisure pursuits serves the same end. Superficially, that end appears to other people involved as the maintenance of a social status superior to that of their 'opposite numbers' in the rest of the concern. Thus, the head of one concern, interpreting the conduct of his research and development engineers in terms of ordinary social differences based on educational achievement, wrote 'The laboratory staff, having graduated from a University, are naturally individuals of a higher intellectual standard than the shop floor personnel, and they have a tendency to consider themselves as superior beings and to act accordingly', and 'designers tend to be young and rather arrogant'.

The cultivation of marks of conduct which could earn recognition as a member of a kind of industrial élite does, of course, have its own reward. But some special significance attaches to the recognition of such conduct as peculiar to industrial scientists, rather than to everybody who enters industry after graduating at a university. We are not, that is, concerned merely with a series of isolated exhibitions of aloofness or snobbery.

This deliberate avoidance is founded on a sense of social incompatibility which contrasts strongly with the consensus about interests and evaluations of people and aims which prevails in the laboratory group. Interviews with members of one large laboratory contained a number of reflections of the contrast. The ease and friendliness which existed between development engineers at all levels of seniority was immediately apparent; Christian names were used between the most senior and junior members, including technicians. They had their own recreational groups and their own informal social activities outside the factory. As for relationships with the factory, 'they are pretty formal; that is, they're confined to functional relationships with other departments. We have nothing to do with the machine shop, for instance. Of course, there have been one or two efforts to get us all together on do's arranged by the firm, but how can you be said to have any sort of social relationships with chaps who are only interested in the dogs and in football matches? It's pretty hard to keep up a conversation at all. . . . Oh yes, a few of the chaps in the lab live near us, and we see quite a bit of each other in the evenings.' This was from a senior member of a design section. A junior member said, with emphasis 'There is no contact with the factory—no personal contact, that is. In the lab we're very happy—sort of happy family relationships. The lab chief must select people on the grounds of getting on with others; they certainly do get on with everybody . . . we have regular meetings with Ministry people and other Government establishments and quite a lot of what you might call social contact with them too.' Another: 'Personally, I keep clear of the factory.'

This last informant went on to give accounts of two occasions on which he had outmanoeuvred a production department and an office department by getting possession of their data surreptitiously and producing results they either could not or would not produce themselves. Naturally, these activities had been conducted on behalf of his laboratory section, and with the approving knowledge of his seniors; no more had been involved than securing the time sheets of a production department to prove a point in dispute about the allocation of hours to a particular project, and the preparation of an estimate for a development and production contract without consulting or informing the factory costs and estimating department, but the effort scored points, for the speaker and those present at the time, not so much in the interests of the concern's effectiveness as a whole as in the perpetual design versus production campaign.

One inevitable consequence, in this concern, of the display of attitudes and of conduct such as we have related was a blockage of traffic in the other direction: from production to design. There was at least one case of a technical problem being referred to a Research Association in London, when the solution might have been found by consulting a few specialists in the laboratories a few yards away from the production department, and under the same roof. So marked was the reluctance among production engineers to consult laboratory engineers about problems incidental to production—to use 'scientific brains'—that top management entertained dark suspicions about 'instructions' being given to stop them. Instructions, had anybody been foolish enough to give them, were entirely unnecessary.

In another concern, production managers reacted more directly and vocally. Objections were raised to the laxity which prevailed among its development engineers over rules which members of production departments had to keep whatever their status—especially punctuality. Production managers objected strongly, as did development engineers, to the proposal that they share a new building. Their present establishments were miles apart, the development engineers in an elegant, pleasantly situated building, surrounded by lawns and gardens and attractively decorated and furnished. The production shops and manager's offices were, by contrast, crowded and makeshift. 'It's often been suggested,' said one production manager, 'that we should make up a party of our people and take them over to see around the labs—good idea to let them see what the firm is doing in that line, and so on. Well, that party hasn't come off and won't. We daren't. What the eye doesn't see, the heart doesn't grieve over.'

The social barrier between development engineers and others made itself felt in every single firm. Even in the concern in which they seemed least obtrusive a senior engineer mentioned an occasion when his team had been spending the morning working very closely with draughtsmen and he had—either absent-mindedly or by way of a gesture of solidarity—joined them at their table for lunch. As he left the canteen he was asked 'Been slumming today?'

STATUS AND THE WORKING ORGANIZATION

No concern, it is safe to say, is without political or social conflicts which generate, or contribute to, manifest inefficiencies of communication within the working organization. What we have been considering is a situation of a special kind, the distinguishing features of which are all aspects of the special status claimed by technologists. The consequences of this situation for the working organization appear in the maintenance and reinforcement of 'linguistic barriers' which occur in the interpretive system.

If the criticisms, typical of those directed against production, recited on pp. 164–5 are reviewed; misreading designs; inability to correct faults in design; inability to refashion a design so as to facilitate production; refusal to accept the need for small quantity production as normal and as a technical challenge; unreadiness to find out correct production methods for special jobs already done in prototype manufacture; four of them may equally well be regarded as pointing to failures in effective communication between design and production. Everything points, in fact, to there being very solid social barriers of status distinction between production and design, and it is possible that what is first of all needed is not supersession of the existing production management by more accomplished engineers so much as a means of overcoming the social inhibitions to effective communication.

In an industrial milieu such as electronics design and manufacture, new problems, unfamiliar procedures, materials, parts, and assemblies are very frequent. This means that individuals are not only constantly having to explore new ground, they are being asked—and having to ask others—to do things which are not written in to their job specifications. They must therefore be continually having to ask others for information or assistance beyond what has been provided before. In the circumstances obtaining in all the concerns we studied, for production people to ask designers what is meant by an unfamiliar feature of a drawing or specification, to ask whether an alternative method of making is possible, and so forth, might well carry too big a risk of a snub or of adding to the body of criticisms of production which had been built up.

The cohesiveness of the laboratory groups and the loyalties and shared interests which transcended differences of status were not only evident to us as observers, they were made much of by our informants, most of whom also pointed out the obvious contrast of their dissociation from the rest of the concern. Occupationally and by ties of interest and occupational status, there were closer associations with civil servants and with members of Government research establishments. Strategies and isolated moves were undertaken not in order to promote individual status-gaining ends, but in the interests of the power and status of the laboratory as a whole. The question which remains, then, is what these interests were, or were conceived to be, and how did they differ from those of the concern as a whole so as to militate against effective organization?

In particular, what distinguishes the difference in status between industrial scientists and the rest of the concern from the differences in status between persons occupying different positions in the management hierarchy, which are regarded as more or less acceptable features of 'the way life is'? Any business concern is rife with invidious distinctions; the management structure exists on the basis of them. What is special about this set of distinctions?

The point has some sociological interest, even if people caught up in these issues have to be concerned more directly with the moral and practical need to take up a position and defend it. The matter is best understood, perhaps, by regarding the working community, which the concern in one sense is, as really a community, into which people are 'born', or into which they migrate, when they take up employment. If we consider the hierarchy of management, supervision and work-people, and the accepted beliefs about it, the characteristic sociological feature is that mobility is possible from lower levels to higher. More, the successful survival of the whole community depends on ensuring that the people with the intellectual and other qualities most desirable for the top positions will eventually occupy them, at whatever level they started. And in order to ensure that this happens, in order to ensure that the 'better people' will themselves try to get to the higher positions, the attractions of monetary rewards, marks of esteem, exclusive rights and privileges are offered for success.

Movement up through the hierarchy is not only normal; it is necessary, and almost universally desired. Yet the laboratory group, while part of the whole situation and claiming at least an equal chance to compete for the highest ranks, claims also the right to a special set of exclusive privileges and marks of esteem based on distinctions acquired before they entered the concern, of 'birthright' superiority, as it were.

The analogy is tenuous, but does serve perhaps to explain the curious readiness of an industrialist to speak of 'physicists' as 'a different kind of animal', of the head of a laboratory to say to a managing director that he 'really can't expect lab people to observe the same rules of punctuality as other people because they just aren't like ordinary people'; of a production manager to say that 'My people are the sort that stick to routines in life; their whole outlook on life and the way they think people should act is different—miles away from the R. and D. lot. They're temperamental, we all know; they look differently at things. Put them together and you'll always get clashes. The only thing is to keep them as far away from each other as possible.'

'Clashes' occur because the special invidious distinctions between the status of individual scientists and other members of a concern are not accepted as 'legitimate', i.e., as in conformity with the expectations and obligations normally attached to membership of the working community.

This is not the whole story. Many entrants to working communities bring with them special social or educational advantages. The second peculiarity of the distinctions we are examining is that they are relatively new in the engineering industry (the indications are that the chemical industry has long accommodated itself to them); also they are presented as an organized affront to accepted conventions. They also threaten the accepted social structure by being attached to a new institutional form—the Research and Development laboratories—which is a contestant for political power in the firm.

STATUS AND POLITICS

In many respects, therefore, the particular style and manner of laboratory workers, and their claims for special treatment and for special privileges, are manifestations of demands which are not compatible with the social and political structure existing within the industry. Both in his demeanour and by individual strategies, the industrial scientist seems to claim:

(i) increased recognition of the importance of technical information as a business resource, and therefore

(ii) increased standing for himself as controlling the sources of such information;

(iii) increased control over the processes of manufacture and over the conduct of the business;

(iv) quasi-élite status within industry as a vitally necessary functionary who is nevertheless not dependent on industry for livelihood or status;

(*v*) recognition of his membership of a group of highly valued members of society free to move occupationally outside industry, and thus of

(*vi*) semi-independence of the authority of management.

The point of central importance, however, is that these aims are not pursued in overt and direct ways, but act as influences on the working operations of people and groups. These are, as we have already said, the only paths of action for political or status-gaining activities within the concern; and when such activities are expressed through decisions or opinions about the structure of the working organization or the allocation of power in the concern, that structure and the actual distribution of power become adjusted to the status structure or political system of the concern rather than to its functional requirements.

CONFLICT AS A CONSEQUENCE OF INSTITUTIONALIZED CHANGE

When changes are manifested not only by the intrusion of new kinds of task and new kinds of resources, or even by the recruitment of 'new kinds of people', but by clothing them in new institutionalized forms, they take on the appearance of a threat to other parts of the existing order, instead of a source of new life.

This is seen most plainly in the unique case encountered in one large English concern of new techniques invading the development laboratory itself. The existing establishment had been employed wholly in electrical work for some decades. In recent years electronics techniques had been applied to one of the two main products of the concern, and development work had been initiated by a small group of electronics engineers imported from another concern in the group. This was set up as a special department with its own leader responsible to the Chief Engineer of the concern to whom the existing chief development engineer was also subordinate. A first equipment was, at the time of the study, already being made in production.

The concern had not reconciled itself to the changes heralded by this new departure. Among senior management there was a tendency to regard its arrival as representing no fundamental change. They thought of electronics as 'essentially a one-off kind of industry' (despite the presence in the group of domestic radio and television receiver manufacture) and as needing a 'different kind of person to make a go of it'. Below this level, electronics represented a threat to the older development and design groups and ancillaries. There was a risk they might become depleted too soon, by transfer and by declining recruitment. The whole works 'had been shattered' by statements said to have been made by the Chief Engineer which put the life of electrical design work at five years, which he had had to call a works meeting in order to revoke. There was bad feeling about promotion. One manager said many senior people had been 'shunted'. Perhaps most significant of all was the feeling among the older engineers that what precedence they once possessed over production managers had now gone. 'It used to be', one senior engineer said, 'that the engineers had precedence over the shops, if ever it came to a stand-up fight. It's not so now. The shops have priority—and the money too.'

Electronics work had come into the firm in the guise of a new, virtually independent organization. No adaptation by the existing organization had been called for, or had been possible. Indeed, the arrangement had been made *so as not to disrupt the existing structure*. Consequently, when the new growth was firmly established, there was only one course open to the former laboratory establishment; the acceptance of extinction, at best after the existing staff had retired or found other jobs.

While it is not feasible for engineers who have spent many years in working with electro-mechanical techniques to turn themselves into electronics engineers, it is feasible that the introduction of the new group into the existing laboratories, had these been flexibly organized, would have rendered the new techniques less of a threat and more of an interesting new departure infusing the whole division with new life.

This, indeed, was the course followed in the production shop handling the new equipment. This single contract—for a large and complex system—was being handled in a corner of the main assembly shop concerned with the manufacture of the electro-mechanical equipment the new system was to replace. The departmental manager, himself nearing the end of his working life, showed obvious enthusiasm for the new venture, and, as much as the development engineers, cheerfully disrupted the formal relationships and procedures, and the sequence of production stages which had been devised to ensure a steady production flow during the decades when the engineering techniques had been virtually stable. With his foreman, charge-hands and fitters, he canvassed ways around production difficulties among themselves and with the development engineers. And in helping to solve the technical manufacturing problems of this particular innovation, he, and his subordinates, were at the same time mastering, or avoiding, the managerial and personal difficulties with which the design laboratories were beset.

POLITICAL CONFLICT

In the Scottish study, the main contestants were the development group, the newcomers, and the pre-existing working organization. The three situations, in different firms, which are recounted below reveal the political issue as a straightforward conflict between the newcomers, in the person of the head of the development group, and the pre-existing organization, represented by the senior management both at his level and above.

(*i*) The first of the three concerns was a recently created subsidiary of a large group. Specializing in heavy electrical equipment, it had begun manufacture with American designs obtained under licence, but had almost immediately sought to create its own design organization. Accordingly the firm had engaged a senior industrial scientist with unusually high technical qualifications and with industrial experience to work on fundamental problems of design. Development work in this firm, therefore, had connotations rather different from those of other places in that successful development work was quite essential to the concern's future as an independent business with its own capacity to compete with others and to expand. This being so, there had been an explicit understanding, when the chief development engineer was brought in, that he should take over an increasing share of the general management, i.e., of production as well as of design. This in fact did not happen and, despite a series of complaints about his position, and subsequent discussions with the directors, he remained in a consultant capacity, supplying technical information at the stage of negotiations for contracts with customers. The situation was resolved by his resigning from the firm.

It is in the nature of things that such a course of events will be regarded by the persons involved as the product either of incompatibilities of character between the principals or of technical or managerial inadequacy. Such experiences can hardly be repeated more than two or three times in the careers of senior managers or senior industrial scientists, so that there is no possibility of their being struck by uniformities about these situations divorced from particular events or individuals. The progressive deterioration of the actual working situation from what was planned or hoped for can only be imputed to the incapacity of people to arrive at and carry out the minor decisions needed to make the initial major decision effective. In brief, what happens is that a plan devised in terms of changing the working organization fails to materialize because factors of status and politics play a determining role, and nobody realizes, or rather, admits, that these are real problems to be dealt with. When these factors intervene, as they must, they are regarded as illegitimate if they are recognized at all, and the person who is thought to be advancing (or resisting) claims for increased power or status is condemned as more concerned for himself than for the job. Usually, of course, there is never any naked exhibition of involvement in the status or political systems, and claims and counter-claims are translated into terms appropriate to the working organization. Yet, underneath the tacit assumption that the working organization is all that it is proper to attend to, that it is the job that counts, there are always actions which reveal that considerations of power and status are fully acknowledged as important elements in the firm and in individual decisions. For example, at a critical point in his endeavours to extend his authority beyond the design department of the concern, the chief development engineer arranged a discussion about the reorganization he proposed with the managing director of the concern and a member of the board of directors of the parent company. At the outset of the discussion he was offered a rise in salary and a new title; later in the discussion, it become apparent that there would be no significant organizational change. A rise in status had been put forward as compensation for refusing a management change which—so it appeared—would have weakened the dominant combination within top management.

(*ii*) The second situation is an almost parallel case. Again the industrial scientist brought in to take charge of the laboratory group had high qualifications and some reputation in the industry. The firm was old-fashioned, and had decided that it must build up technical resources which would render it independent of obsolescent products and methods.

'The understanding was,' said the head of the laboratory, 'that I was needed by the firm, I was to take over their whole technical side and be responsible for technical development. My view now (after two years) is that this has been a wholly unsuccessful venture as far as I am concerned—or the firm. . . . Everything I've put up has been blocked, it's met with deaf ears. As far as I can see, the firm just hasn't been able to see what to make of me.

'I've had several propositions which I've put up to the firm; none of them has been accepted. It's been their view that whether or not such and such a development takes place is just not my affair; this is the manufacturer's concern, and similarly the question of how something is to be developed or produced is regarded as not my affair either. So far as I can see, what they expect the lab to do is to hand over a blueprint of a new product, which the manufacturing side, the works manager, could put into production almost straight away. . . . They have no conception of what development work in this industry involves . . . they have constantly refused to put aside a certain amount of capital each year for development purposes.'

The basic political issue was again the expansion of control by the laboratory group into sectors of the concern previously under the exclusive control of production and higher management. It had, however, been exacerbated by actions of the research manager; soon after his arrival, he had submitted a report on the equipment and operating methods of the production departments which had been extremely critical. From this beginning relationships between the laboratory and the rest of the concern appear to have flowed in ever narrowing channels. From the laboratory came reports, criticisms of faulty techniques, demands for new equipment, proposals for intervening in the running of the manufacturing process in order to prove laboratory trials, all of which were construed as arrogant and irresponsible. From the other side came refusals to embark on any change or new product except one which would be a guaranteed success or would cost nothing; this was construed as obscurantism and meanness.

The research manager had recruited a small team of industrial scientists, including some three or four graduates, but the effective work of the group was confined to technical control of the product at various stages, including the testing of raw materials and products. This had, of course, been done before his appointment and the build-up of the laboratory group. It was now done rather better.

Three years after his appointment the research manager resigned.

The situation had reached a point, before the end, at which he could say that he had 'not spoken one word for over a year' to the managing director, whose office was next to his. It was the same managing director who, after being on friendly terms with him for years, had invited the research manager to join the firm.

Here, as elsewhere, the only factors contributing to the situation which were at all visible to the people directly involved were personal. And, as the situation confronting each person grew more intractable, so his own aims and their frustration became more charged with emotion; so that in the end, organizational defects had in fact been translated into personal terms.

(*iii*) Treating the industrial scientist as a slot machine capable of producing on his own, and without any connexion with the manufacturing side, innovations ready for production, is a necessary corollary to isolating him. The Research and Technical Manager in a third firm said that his directors 'think I'm a conjurer. They think that by hiring me they ought to get it out of a hat.'

This third concern, linked with the shipbuilding industry, was also old-established. During the war the firm, in common with the rest of the shipbuilding industry, had been working at full pressure. In 1945, with a greatly increased production potential and the probability of a rapid fall-off in demand from its existing market, it prospected both for new products among recent developments in its own sector of industry and for new markets. Having made a first successful venture in the supply of new type fittings for the oil industry, the directors decided to prospect further, and to engage an industrial scientist to do exploratory work among recent innovations and among the possible markets for them.

No very considerable new departure resulted from this for some time, but the firm's existing range of materials and processes was extended on the basis of technical information collected and collated by the new Research and Technical Manager. It was in this way that he began to be fitted into—and accept—the consultant role which he later found irksome. Furthermore, since a great deal of this work at this time consisted in technical sales—i.e., study of the more difficult requirements of possible customers and the suggestion of ways of meeting them which the firm might undertake—followed up by technical service, he appeared to be fitting into a sales role. In the role of technical salesmen and consultant thus devised for him, the promptness needed if technical answers were to be of use to the customer or the firm forbade any growth of development work of a longer-term nature; the firm was doing well enough without any sizeable capital outlay on equipment or on work which could not be charged against a specific order from a customer.

At this point a major opportunity appeared, in plastic material which could be substituted for the metal traditional to the industry. There were many technical advantages in the new product, a low cost, and the market for it was one very familiar to the firm. The only big problems were of manufacture—of producing sufficient quantities —and of versatility in matching different customers' needs.

The usual situation now revealed itself, although the symptoms were, again as usual, idiosyncratic. The industrial scientist was appointed manager of the new department, thus switching part of his functional role from a position external to the manufacturing side to one in which he would be subordinate to the production manager of the works. This was in addition to the functions which had already grown up, and 'I have never had a laboratory, laboratory equipment, test equipment or any technical assistance in terms of personnel. I have had to develop this on theory, I've got to dream up the right conditions, and the right temperatures . . . go straight out and try it on the production machines.' Not only was development work proper virtually impossible, there was also no expansion of production comparable with that of demand. This, the Research and Technical Manager thought, was largely owing to the absence of adequate costing. The management had always relied on the estimating procedure which had been quite good enough in pre-war days of minimal changes in cost factors, little technological change, and a familiar, single, and traditional market. He firmly believed that the profits being made on his new product were being used to offset losses made in the main production shops; this also applied to the development charge normally added to the selling price of the new product. Feeling that there was an all-round resistance to the growth of his own functions, to the expansion of his production department at an adequate rate, to the legitimation of the powers of control he possessed over the effective matching of the firm's manufacturing resources and market needs, he resigned.

All these three cases display situations of conflict essentially political in character and similar in pattern. In every case, an industrial scientist was recruited in order to promote the growth of the firm, or to improve its chances of survival; the initial status of the newcomer was roughly equivalent in each firm; his organization chart position, 'sticking out in the air at the end of the line', as one put it, of departmental and functional managers immediately below the managing director's level, was identical. The role alloted to the newcomer was external to the existing structure. His powers were not permitted at any point to infringe on the control previously exercised by other members of management.

In other words, the industrial scientist was fitted into a consultant advisory role; decision-making and control over the operations of the firm, including any that might be undertaken on the basis of the expertise contributed by him, were reserved for members of the former executive hierarchy. Although one did complain that he had no laboratory resources, this was a fairly minor dissatisfaction; there was no mention of any such limitation by the others, and one was fairly clearly of the opinion that his specialist staff and resources were grossly in excess of the needs they were fulfilling. There was apparently no control exercised over the directions of technical exploration they chose. There was no indication that the technical discoveries and innovations they brought forward were impracticable or irrelevant so far as the firm's market or production methods were concerned. The issues were almost solely concerned with their claims for greater control over the activities and resources of the firm, and the resistance to these claims by the rest of the management. In adding special technical skills and sources of information to the firm, that is, no allowance was made for altering the fields of control exercised by different members and groups in the firm's structure—for bringing the new functions into management.

Not all political conflicts are lethal. Conclusions such as those which were reached in the three firms which have figured in the last few pages are seldom experienced, partly because the growth of one part of a concern need not necessarily involve the total extinction of another, partly because there is rarely such a sharp division made between new and old. But political conflicts do appear out of situations in which changing circumstances constitute a threat to existing parts of the working community. This happens when the new circumstances themselves are institutionalized—i.e., when a special person or organized group of persons is designated as the agent of change, the existing organization being relegated to the role of spectator or patient.

Such a role is rarely, if ever, accepted. The ensuing conflict is one which may be directed towards establishing one's own right to existence, or denying it to the other side. But the outcome, whatever the fate of the two sides, is to shape the organization of the concern according to the shifts and expedients employed by the contestants.

POLITICS AND THE SHAPING OF THE WORKING ORGANIZATION

Perhaps the most interesting and enlightening political campaign encountered in these studies was that successfully conducted by a Development Laboratory to establish its right to separate autonomous existence between a Research Laboratory and the Production plant. All these were situated in different towns, although the intention had been to put the Development Laboratory and factory on the same site. All the work on which Development and Production were engaged had begun in the Research Laboratory; the Development Laboratory and, some four years later, the factory, had been newly created.

While the distance between the two laboratories took some time to cross, it was also seen as an advantage by development engineers. There was little doubt in most of their minds that separation made much easier the independent critical appraisal of the products of Research and the ability to decide on redeveloping. It was said, for example, that Research began by believing that the Development Laboratory would merely take over a design that they had produced, check the tolerances, see that it could be engineered, make a prototype, and send it for manufacture. But this 'doesn't often happen; more than likely it has to be entirely redeveloped. So Research take on study contracts, and we produce something that will sell and meet specifications—something that can be made as economically as possible, will continue to work, and so on.' Even when remoteness was not seen as a positive advantage, the drawbacks were regarded as negligible; 'we speak the same language and there's a good deal of frank talking on visits. . . . We know what they're doing very early on.' Information had to reach a certain level of importance before it became worth while going to the trouble of communicating even by telephone; 'this means that there exists something about which a practical discussion can take place'. Separation also tended to face Research with requirements for specific jobs of work; 'instead of an idea just growing and growing'. With Research and Development in the same establishment 'it is often impossible to get a design cleared up stage by stage; the tendency is to say "Hang on, we've just got a new idea," and this tendency can work right through the process'. Lastly, separation and the fresh scrutiny which a project could receive when it passed from research to development might 'prevent rogue ideas going right through without being stopped'.

Many of these statements reflect the growth of a quasi-inspection function. This function had been turned to effective political use. 'Relationships with Research weren't at all good three years ago. They couldn't understand why their brilliant circuits were criticized. Now they know by experience that they can go wrong.' A number of informants in the Development Laboratory recalled difficulties in co-operation with the Research Laboratory during the first years, because of the unwillingness of Research to accept the fact that development groups had a significant part to play. The improvement which came about was ascribed by development engineers largely to the difficulties which occurred in the manufacture of prototypes. Research became willing to let Development sort some of these troubles out. Status distinctions between the two groups also became less obtrusive. 'Most of the troublesome people have left—you really can't have research people who think they're God Almighty.'

A second strategy again linked organizational needs with the acquisition by development engineers of complementary and equal status. This was a system of pairing between the 'opposite numbers' working in either establishment on the same equipment, or technical item. Pairing extended down through both establishments and worked not only through informal contacts but through the membership of all progress meetings. Each establishment provided the chairman on alternate occasions, and both members of the pair from Research and from Development concerned in any subject were present at the meeting.

Over a number of years the Development Laboratory struggled to insert itself between Research and Production until, in 1957, a balance was struck and given titular recognition in an identical name for both establishments (invidious professional distinctions between Research and Development being dropped) and in the appointment of the heads of both as Joint Managers.

The Development Laboratory viewed altogether differently the almost parallel situation between itself and the production factory, which had been created later, and had also been sited a good distance away. This factory had begun with dreams of being 'a genuine production factory' and still strove for autonomy. Production pressed its claims for autonomy by demanding clear designs and specifications and by claiming the job of translating these into manufactured articles without Development retaining responsibility. These claims were totally rejected by Development.

Many of the moves and ideas which figured in the account by development engineers of their group's struggle for autonomy and equality were repeated, with the significance reversed, in the account of relationships between Development and Production. However salutary the influence exerted by a separate and remote Development laboratory on Research, having Production so far away was 'thoroughly bad'. It was much better, said the development engineers, for the designer to 'live with the job'—to have design responsibility for equipment throughout production. Only by having development on the spot with production was it feasible to maintain this kind of control; otherwise there was a tendency for a handover of responsibility to take place. 'What happens is that you're constantly getting unsuspected faults arising from characteristics which you didn't think important in the design. If you got to hear of these through a . . . production person to whom the design was handed over in the dim past, then instead of being a design problem, it's an annoyance caused by that particular person; you thought you were finished with that job and you're on to something else now.'

The whole two-sided situation is neatly expressive of the way in which political pressure is rationalized in terms of organization. The reversals of argument were taken even to the point of explicitly suppressing pair relationships (so valuable in the build-up *vis-à-vis* Research!) which had grown up between development engineers and production people. 'We've set up a production liaison chap to act as a channel for communication. You can get all kinds of private arrangements between production and development people . . . personal knowledge is no good to anyone.' The overt expression of the Development laboratory's campaign for power and status equal to Research and for control over Production was made in terms of the exigencies of organization and had been by moves made in the interests of more effective working. The structure of relationships between the three establishments was, in turn, the creation of the successful manipulation by the Development Laboratory of its intermediate position to establish first a necessary and finally a dominant role in the whole tripartite system.

NEW POLITICAL ALIGNMENTS

Among the large, well-established English firms, the biggest and most prevalent of political divisions was that between sales and development-design groups. The issues were at one level clear-cut, and had to do with the relative contribution to the commercial success of the firm made by either side and the right thereby to a greater or less share of resources and capital for expansion. At this level, as in the case of the design-production conflict, the campaigns were conducted not in overt demands for more power but in operational terms; in attempts, or threats, to 'shop' the other side, i.e., to expose their incompetence, or in criticism directed to that end. 'Our sales people spend all their time looking sideways' (i.e., at what competitors are doing) 'rather than forwards. They are resistant to new ideas.' 'They think exclusively in terms of turnover, and not of profits or of the future.' 'They ought to be able to go further than just to say what they want is what the other chap is selling, only cheaper and better. They ought to be able at least to say what they want in terms of the lowest possible price—but they can't do this.'

Much of this hostility was newly engendered by the shift from government work to commercial and the consequent loss of the development side's sales function, and the build-up of the sales organization (see Chapter 4). 'We used to work for our customers (Ministries) direct—they were our masters, really. . . . Now Head Office are our masters.'

Hostilities seemed to be particularly keen in the companies in which the supremacy of the sales function had been registered by the creation of product divisions, with managers primarily involved in sales assuming all or some responsibility for development policy. A third factor was added if there was some geographical separation between design and sales, as was often the case.

'Control is much more firmly at head office than it was—this goes for development money and the sort of things we do development work on. They have complete control of sales, of course.'

'You get a lot of trouble now with sales. Things which could be sorted out in two minutes if somebody was in the same building can now blow up into major troubles . . . you get delays, and the sales people, because they have no real knowledge of what is happening, or who the people are, they get to blaming people, whereas they would understand what was wrong if they were in the same building.'

What was in the domestic radio manufacturer's case merely a desire to get sales to commit themselves more to specifications for new designs, and to enthuse more about bright ideas put up by design, was in other cases a much more serious resentment of a sales hegemony which had recently made itself felt as a threat. The increase in the size and importance of sales was something patently related to changes in the general development of the industry and its market and to changes in government demand. The sales function of development engineers was still a reality, even in the firms most dominated by Sales divisions, and the essential tasks of development work remained. If anything, the challenge of increased competition in obtaining government work and the scope for initiative in new kinds of commercial projects provided a more stimulating setting for the engineer than he had had for many years. And development and sales were members of the same enterprise; the future of each was to a very large extent dependent on the success and growth of the other's activities. Yet bad feeling, complaints of misunderstanding or incapacity, and lack of communication between the two were one of the most usual features of the firms surveyed, and were in some firms, large and small, a serious management problem.

NEW-STYLE CONFLICTS

No firm, it is as well to repeat, was without some serious political conflict dividing it. It need not, of course, exist between departments, and perhaps the most interesting of all the firms were those in which political divisions cut across the functional divisions. In many of the Scottish cases we have quoted, production had obviously stood for the previous order of things, and the laboratory groups represented the new dispensation. In one English company the same conflict of old and new visibly underlay the political split within a large design department (pp. 192-3). The issue there lay between those working in technical fields which had remained basically unchanged for a generation and those recruited for new electronic work, which, it was expected, would eventually replace the older techniques. But, typically, the main division observable in the English study was between sales and development. On a closer view also, it seemed that the sales-development conflicts had become noticeably acute in the last year or two, with the increase of sales effort in the new commercial market and the shrinking or disappearance of the implicit sales function discharged by design engineers in the government market (see pp. 56–65).

What is striking about all these kinds of political conflict is that they were allied to the appearance of a new group, or the rapidly enhanced importance of an existing group. This threatened the power, influence, and prestige of the formerly dominant groups. The political issue between sales and development was moreover particularly acute where the growth of sales had been accompanied by reorganization into product divisions, in which the leading role fell to sales.

Thus the main political issues in most English firms seemed to arise from the resistance of development and design engineers—the professional innovators—to an innovating change, much as production people in the Scottish firms had resisted development engineers as the instruments of change. But political issues arose not from the fact of change itself, but from the identification of change with one section of the concern, whose new role and expansion threatened the power and standing of other sections which were being treated as though they were unaffected by the new dispensation and could be left to accommodate themselves passively to it.

PART THREE

Direction and the Shaping of Management Conduct

CHAPTER 10

The Men at the Top

The analytical scheme developed in Chapter 6 suggested that the approximation of a working organization to mechanistic or to organic form was determined by the operation of three 'variables'. The effect of the first of these,—the rate of technical and market change—was discussed in Part I and in Chapter 5. The second—the strength of personal commitments to the improvement or defence of status or power —formed the main subject of Part II. The third variable concerns the extent to which the managing director can interpret the technical and commercial situation, and can adapt the working organization and elicit the individual commitment to it for which the situation calls. This is the subject of the remainder of this book, although, of course, it has made previous appearance in many connexions (v. esp. Chapters 4 and 7).

The two aspects of the director's task which we have specified as distinctive are related to the two dimensions of the working situation of all subordinate members of the concern: the 'framework of organization' and the 'framework of decision'. In this, we follow C. I. Barnard[45] and H. A. Simon (see quotation, pp. 115–16) who have represented the 'physiology' of organization as a process of composite decision. In a mechanistic system decisions at lower levels are taken within the framework of decisions at higher levels. Also, all that a superior in fact does is to define the context, the conditions, the premises and the grounds within and on which subordinates act. In organic systems, subordination becomes less important as determining active and passive roles in this process; everybody in the system has to work out his own actions within a series of temporary 'frameworks of decision' set by people around him.

Nevertheless, seniority does count in every business concern, whatever its management system. As we have said (p. 122) the 'lead' is taken more frequently by those best equipped with local and technical information. Moreover, there is a tendency for every working organization to revert to a mechanistic system, since this persists as the dominant model of organization.

Seniority, the more frequent possession of superior local and technical information, is easily converted into superordination, which is in operational terms no more than the formalization of seniority; it becomes an assumption of the people involved that better information always and necessarily attaches to the senior by virtue of his position. Such an assumption is more readily made by seniors, whose position is reinforced and whose incumbency of it is made more secure; and it is the more readily accorded by juniors when their selection and promotion is dependent not on the consensus of all around them but on the view taken of them by their senior.

Secondly, as we have tried to show, a mechanistic system is more economical of the individual's effort. Commitments to the working organization are more prescribed the closer the approximation to mechanistic form. The tendency is for most individuals to oppose extending such commitments and to try to reduce them, and thus to exert pressure towards a mechanistic system. If conditions are stable, this means that overall economy in human resources may be effected. If conditions are unstable, a mechanistic system becomes extravagant in numbers of persons employed (cf. pp. 168–71 on the growth of intermediaries and interpreters) each with his limited commitment to the working organization.

Visible power, seniority, and the pressure towards limiting commitments and mechanizing the working organization all focus on the managing director. In the last resort, in the modern industrial concern, it is for him to define the situation—the organizational structure and degree of commitment—of all others.

THE MANAGING DIRECTOR'S ROLE

For the managers contending with the difficulties and waging the conflicts described in previous chapters, the clue to the situation often lay in the character of the head of the concern. Only a long time after he has gone, or across a safe barrier of space, is he seen by his juniors as the creature of circumstances as well as their creator. The one constant element of all the studies of the twenty concerns was the extraordinary importance ascribed to the personal qualities of the managing director or general manager of the plant. In many firms, almost every interview would contain reference to the 'outstanding personality', the 'flair', the 'wisdom', the 'tremendous personal courage', even the 'genius' of the managing director, and the all-important part he had played in the success of the firm. Even in those concerns for which the auspices were less favourable, many managers would be at pains to stress the 'niceness' of their chief 'as a person'.

It may, of course, have been thought good policy to make such remarks even to an outsider. Yet the experience was so frequent, and the contrast between what was said about the head of the firm and the critical appraisal of colleagues, even when they were credited with outstanding abilities, was so pronounced, that the cynical explanation can only be partial, at best. And even if it is granted that such remarks were merely a precaution or a propitiatory ritual, they exemplify with sufficient force the pervasiveness of the influence cast by the managing director over the conduct of members of the concern, an influence which extends far beyond the occasions when he actually and overtly exerts the authority he possesses.

The head of the concern stands for the concern and its relative successes—he symbolizes or personifies it. The management of the concern is also a career system, and the man standing on the topmost rung has to serve as a showcase for the characteristics which must be attributed to the person who is by definition the most successful, or their absence must be condoned by the presence of equally highly valued personal qualities.

So, while there is a regular succession of increments of authority and higher status positions as one moves up the management hierarchy, the topmost position differs from the others by more than a matter of degree. As the ultimate controller of direction, he sets the goals the whole concern is to achieve, and so sets the parameters of the task and activities of groups and persons. As the 'patron' of the concern, he specifies—by example, by adjudication, by remonstrance, by permissive silence, by approval—the measure of privilege and rights attached to lower positions. The system and structure of management are both determined largely by him. Above all, he is the ultimate authority for appointment and promotion.

The head of a concern, therefore, exerts a powerful influence over its members in their conduct in many ways apart from the actual and overt authority he possesses. In addition, he occupies a socially isolated position at the top of the management hierarchy.

'Social isolation' comprehends more than the 'loneliness of command', the familiar notion of the burden of responsibility which is carried by the person to whom is ascribed ultimate authority in any organization. It has to do rather with the social peculiarities of the position. People in working communities in our society normally have to deal with others who are subordinate, equal, or superior in social standing and power. In doing so, they are continually testing out the answers they have themselves arrived at to such all-important questions as 'Am I a likeable person?', 'Am I doing my job well?', 'Did I act on such and such an occasion intelligently, or unjustly, or dishonestly, or with enough warmth, or with enough detachment?' In the relatively crowded milieux of our occupational lives, we are presented with plenty of hints, clues, and even downright statements from equals and superiors which we can trust as evidence of fairly disinterested judgement on these matters. We do usually trust them, at least when they are favourable; it is the unfavourable judgements which we can often dismiss as unreliable, because of the element of competitiveness which enters into all our dealings with colleagues and superiors.

As soon as a man reaches the top position in the hierarchy of a working organization, this normal situation is reversed. He enters what is in many respects an unreal world, one in which all responses to his actions and to himself are filtered through the knowledge that he is in supreme command, and in a position to control the careers and occupational lives of all the members of the organization. People below him have to be—and have much experience and training to help them be—circumspect in their dealings with him. They have as a matter of unthinking habit to control not only their utterances but also the manner in which they are delivered, their non-verbal conduct. For they have first to present themselves in the best light and secondly to display to him the precise responses, hints, and clues which will give him the kind of reassurance about himself and his conduct which they think he needs at that moment. None of this should be identified with the cruder forms of displayed subservience, which, indeed, is characterized as such only when it is failing in its purpose. Disagreement and even, in certain circumstances, criticism is normally possible if the timing and style are correctly judged; indeed, they may convey a deeper reassurance to the head of the concern by appearing to discard some essential elements of his isolation.

The managing director may seek either to exploit his position or to escape from it. In doing either, he may act in accordance with ideas prevailing in the concern about what a man in his position can legitimately do, or he may not. This gives four kinds of relationship with the members of the concern, each of which promotes a specific array of responses, and so acts as a chief determinant of the management system. These four kinds of relationship, which are also four ways of dealing with the fact of isolation, are elucidated below.

(i) Exploitation of the managing director's position of isolation and supremacy which is accepted as legitimate can approximate very closely to 'natural' (charismatic) leadership. This was visible in one concern in many instances, but was particularly apparent in the manner in which changes in the management structure were carried out. They seemed to be decided and acted upon with formidable speed, and to emerge from the managing director's personal appraisal of situations.

The task of analysing and changing the system was seen as implicit in the managing director's role. People in the organization sanctioned this because of his identification with the company's success and because he acted as a focus for the general orientation of the system towards swift, imaginative, and intelligent appreciation of user needs in relation to the firm's capacities. A passage in one interview with the head of a recently created specialist laboratory illustrates the way in which structural alterations were seen to occur, and the way in which the managing director's authority derived from his identification with the success of the firm, and with the whole active working of the firm.

Interviewer: 'When this decision was made three months ago to hive off (from the research laboratory), who would have been involved in that decision?'

X: 'Oh, I think that was a personal decision entirely of the Managing Director. The decision was made within half an hour and I was appointed to the job an hour after that.'

Interviewer: 'You mean it was a kind of thunderbolt?'

X: 'Absolutely.'

Interviewer: 'Nobody expected it?'

X: 'No. In fact I know what happened. There was one meeting where the Managing Director decided the situation wasn't working. At the end of the meeting he decided that something had got to be done, and within half an hour after that he called me into his office and asked me if I would take on the job.'

Interviewer: 'Does this follow a normal pattern?'

X: 'You mean the decision taking? .... I think it is one of the reasons why this company has grown as fast as it has, that we have a Managing Director of tremendous personal courage and, once he has made up his mind, things happen, almost within an hour. I think it is true to say that he doesn't consult many people before making decisions ....'

Interviewer: 'But he has a lot of information?'

X: 'Oh, a tremendous amount. Probably more than, well, the majority of men I've come in contact with before.'

Interviewer: 'How does he get that information?'

X: 'By coming and digging into the job. And it's a principle that runs right through the company that he expects when necessary to go and tackle any job or any detail that comes out, and he expects his staff to do likewise.'

As a recent beneficiary of the system, X may well have been prejudiced in its favour. Yet his remarks, quite apart from the many pointers in the same direction given by interviews with his colleagues, indicate the existence of a mutually sustaining set of beliefs existing between the managing director and his subordinates about the former's sovereign powers and rights. These beliefs not only sanctioned his acting in a manner which seems patently autocratic but, by justifying his so doing, protected the self-esteem of his subordinates. Indeed, far from feeling diminished by the speed, assurance, and self-reliance of his conduct, they derived confidence in themselves and in the tasks set for them from their belief in the sureness of judgement which had matched them together.

Charismatic leadership of this kind has served as an archetypal model for all leadership roles. It has often been regarded as the one essential quality demanded of the person at the head of any organization or group. Its existence has been regarded, too, as a substantial justification for organization according to hierarchic forms and the concentration of power in the person at the top. Yet any concern operating on such a system of beliefs, however successful and forceful the leader, is ill-equipped to survive the end of the careers of the pioneer group. When such beliefs become weakened or extinct, the commitments created have to be sanctioned first by explicit principles of loyalty to the boss, or to the firm, and then by routines of subordination. The creation, largely, of the shared experiences and successes of wartime and of the early days of the concern in the case which figures here, the relationships between the chief and his immediate entourage —and the role it sustained him in—were historically unrepeatable; indeed, newcomers to the present management might find accepting the beliefs as difficult as sharing the intimacies of the original group. Mistakes in direction and in selection could be peculiarly embarrassing to correct, or might even remain undetected.

(ii) When the beliefs of the managing director in his capacities are not shared by his subordinates, attempts to sustain the role of charismatic or even of authoritarian leader may in fact result in over-playing or under-playing. Over-playing the role is the more familiar of these two illegitimate uses of the characteristics of the position. It leads to a more complete social isolation. A sudden change of policy, the upsetting of promotion prospects by importing a senior manager without consultation, wayward changes of intention during a meeting, outbursts of temper, dismissals or promotions on what appear to subordinates as insubstantial grounds may, when they intrude themselves into the traffic of conversations and meetings between the managing director and his subordinates, prompt the same, equable, non-committal responses and comments or even the same compliance or applause that other more acceptable or even successful strokes in the past have won. The perpetual encounter, universally within the concern, with responses which are either fabricated or blank sets an increasing distance between the man at the top and his subordinates. His situation can, in fact, approximate to that of the known psychopath who is confronted with an unending and unbroken series of interactions in which his opposites are playing him false, withholding normal reaction to his conduct, and substituting any convenient response which will soothe or placate him and allow them to escape. So far as the organization itself is concerned, the positive aspect of the response is the building-up of a tacitly or explicitly collusive understanding among subordinates; which may develop into team-play against the managing director. This requires a substantial and comprehensive network of loyalties, from which any persons have to be excluded who are suspected of buying favour by acting as informants to the managing director.

Under-exploiting the role is perhaps less familiar, at least in this industry, and was observable in only one case. It consists in a partial abdication from the role. The head of the concern, in this instance, was said to be 'involved' in the big decisions about financial and commercial policy and to interest himself in people and things which had meant much to the company in the past; he was accorded very high prestige because of the 'personal achievement' which the history of the firm, founded by him, represented. But the centre of gravity, so far as general matters of direction and organization were concerned, was on the main board, among his immediate subordinates.

In this case, the collusive arrangements between subordinates, while necessary in order to stabilize the division of power between themselves and their superior, took on an additional function. The ties of loyalty derived not so much from mutual self-protection as from a kind of tacit non-aggressive understanding. In order to ensure that no single member of the ruling group on the main board extended his influence at the expense of the others, a peculiar formal structure was arrived at. A hierarchy of committees on which they all sat *ex officio* enabled them to make or influence executive decisions or recommendations at two or three levels below their own, *and* to review them subsequently. The degree of central control exercised over all aspects of the concern's activities was therefore abnormally high.

(iii) The managing director, however, may escape from most of the circumstances of his social isolation, although since such escape always requires deliberate acts of choice, it does not necessarily affect his ultimate seniority and responsibility. Legitimate escape from isolation was evident in one concern, in which there was a conscious and deliberate effort by the managing director to involve other directors and subordinates along with himself in every decision of a directional or organizational kind. He put it in these words: 'What has emerged in our way of trying to run this business is that although we have believed in the need for committee decisions, we have found that we have had to try and evolve a system of having committees to voice opinions, leaving no doubt that some joker has to go off and take action having heard what everybody else says. This is more sensible if he has said his piece. Even then, the decision may not be the one you'd made. . . .In our set-up we are feeling our way towards the new types of organizational structure which are going to be required—they are based much more on discussion than they were.'

In fact, the system·involved something more than perpetual committees. The essential feature again lies in the character of the working relationship between the head of the concern and his immediate subordinates. For a clearer demonstration of this we may refer to a rather special instance of legitimate escape from isolation, one in which the 'group' into which the managing director dissolved his isolated role consisted merely of two people—himself and the Technical Director. They acted in close concert, spending some hours of almost every day in consultation together, bringing complementary commercial, technical, and manufacturing skills to bear on the issues confronting the direction of the business. Even this single intimate relationship, however, preserved the head of the firm from isolation. It also set a model throughout management for planning and making decisions in the light of consultation, and for continually re-defining responsibilities in accordance with what the situation needed. A number of informants insisted on the major importance of the Managing Director's manner. Not only did he display a friendly interest in anybody he met, whatever their rank in the firm—he could quickly become involved in a working relationship with them, for however brief a time. In fact, the intimate alliance between him and the Technical Director, besides providing a stable element in the whole system, served as a demonstration of how to make a working relationship really work.

The working relationships between the head of a concern and his subordinates are the observable aspect of the way in which he deals with the fact of his isolation. In no case, of course, can he truly escape it. This was put with all the force of an Irish paradox by a subordinate of the first of the two managing directors cited in this section. He said, in an attempt—given independently of his senior's—to explain the rationale of the system, that 'You get the man at the top quite rightly saying "I'm not going to be the man at the top, otherwise there's going to be a complete bottleneck".' So also, the same managing director, in attacking the surreptitious growth of minor privileges as invidious marks of status, found himself designated as their unshakeable foundation.

'It's incredibly difficult to get rid of them altogether. Sometimes in trying to get rid of them you find these very measures being turned into a new system of status designations. The kind of trivial thing— we're always having trouble about parking spaces. So we had them allocated to chaps with cars—all done equitably, and new boys to wait their turn. But in a matter of weeks it all starts up again, and when I remonstrate, I'm told "It's all right for you—nobody would ever dare pinch *your* space".'

In describing a mode of playing the role of managing director as 'escaping from' its isolation, therefore, the phrase should be understood in terms of a perpetual intention and endeavour rather than final achievement. The difference between this and the first two modes is nevertheless material. It may be roughly indicated by the distinction in common usage between 'directing' and 'leading'. In terms of the top management organization, the difference lies in the location of indeterminacy in the situation of the individual subordinate. Under 'direction', indeterminacy is largely confined to the behaviour of the man at the top. He may propose, or dispose, 'without consulting anybody'. One measure of his superlative qualities as head of the concern is, indeed, his unpredictability (although another may also be the subsequent success of his decisions) as showing his confidence in his capacity for thinking farther and faster than his subordinates. So, even though believing in his capacities and stimulated by the success of the enterprise under his control, subordinates are compelled to devote attention to reading his intentions and analysing the factors affecting them, obscure though they are. Under 'leadership', indeterminacy attaches not to the person of the man at the top, but to the situation the subordinate finds himself in, along with his superior and his colleagues and the firm; decisions do not arrive like 'thunderbolts' from above; the only way out of the difficulties of reaching decisions in the most indeterminate of situations is by consulting his superior and his colleagues.

What should be apparent is that both systems equally originate in a line of conduct pursued by the head of the concern, both are equally forms of organization, both equally dependent on his powers and the use he makes of them. For their success, also, both are equally dependent on the extent to which both superior and subordinate share a common system of beliefs about the way the managing director should perform in his role. Because only in so far as the system operates in a context of trust can the energies of subordinates and chief alike be

directed to solving the problems presented by the situation of the concern, rather than the problems of their relationships with each other.

(iv) When a manager makes a confidant of someone who is not
accorded the rank of deputy of next-in-line, or 'plays off' individuals
and groups against each other in order to weaken potential competition
for his position or pressure from below, he may be said to seek an illegitimate escape from isolation. If one took all comments made in
interviews at their face value, this strategy would be regarded as the
most frequent of all four. Some discounting of such statements is necessary for a number of reasons. Allegations of favouritism and of playing-
off are useful as a means of saving one's self-esteem on occasions when
a superior has shown a preference for others, more attention than one
expects to others' affairs, or less to one's own, or has offered criticism
or shown disfavour. Also, most managers react strongly and immediately to the slightest suspicion in their superior's conduct of such
tendencies.

These sensitivities are aroused because such conduct by the managing
director exploits the ambiguities of the positions occupied by members
of a concern. They are at one and the same time co-operators in the
common enterprise of the concern and rivals for the rewards of successful competition with each other; just as the management hierarchy is
at the same time a single control system and a career ladder. So far as
the working organization is concerned, managers are co-operators and
the hierarchy is a control system. It is assumed that this is the aspect
which the head of the concern wishes to be dominant. If, however, he
becomes preoccupied with the weaknesses in his own position, some
of which may be unavoidable, such as being less technically expert
than his juniors, he may feel driven to making use of the competitiveness of his subordinates and to exploiting the absolute control he has
over the success ladder. But he may also do this in an attempt to buy
relief from the rivalries themselves.

At all stages of the management hierarchy, some feelings of insecurity
and self-doubt enter into individuals' occupancy of their several positions. Reassurance of some kind, even spurious, becomes an absolute
necessity at times. For the most part, such reassurance is gained through
gossip. In gossip, judgements are passed on to other people, mostly of
a depreciatory kind. In this way one can gain a temporary 'fix' of one's
own standing or prestige; one can also detect what styles of conduct
and directions of activity win approval from one's peers. Also, 'Gossip

offers the guarantee that because one is united with at least one other in judging A to be deficient in technical knowledge, B to have made a gaffe, C to be sycophantic, D to spend too much time chatting in the canteen, the speaker and his hearer—compared with these others—are at least free from such faults. In gossip, speaker's and audience's status claims are underwritten, relative to these being discussed.'[1]

Clearly the greatest possible reassurance is obtained when gossip is exchanged with a managing director. One managing director remarked that he seemed to spend most of his time 'keeping his departmental and lab chiefs happy'. This appeared to involve long gossipy chats with one or other of them, in which a fairly cosy intimacy could develop. The fact that he had to do so with each in turn, and reject any special intimacy in meetings with more than one of them, kept the jealousies and suspicions he was contending with well stoked. Because, in such a group, it is not long before it is apparent to those who are privileged to enter into such relationships that they may also well form the subjects of gossip on other occasions.

Whether he allows himself to be captured by one or other careerist subordinate or by one or other caucus seeking a political victory, or 'keeps the peace' (his own peace) by making diplomatic use of confidential gossip, the managing director is endeavouring to escape from the isolated position which he must occupy before he can insist on the cooperative aspect of each managerial role and the operation of the management hierarchy as a control system. The difficulties of relationships among the top management of which so many managing directors complain are 'presenting problems' of symptomatic rather than fundamental nature. The essential problem is the unbearable isolation of his own position.

THE MANAGING DIRECTOR AND THE WORKING ORGANIZATION —SUCCESSION PROBLEMS

The management of his own role, then, can be of critical importance in the managing director's definition of the situation for the rest of the concern. The relationship between these two aspects of his conduct is particularly clearly revealed when they are given a highly personal significance in the problem of the choice of his successor.

The origins of major political conflict have been identified as change when it appears within an organization under the guise of a separately constituted institution which is either new or is given a recognizably new role including new status or new powers. But change in the future, containing the threat of such manifestations, may also produce conflict in the present. Years before a managing director, or perhaps a whole generation of top management, is due to retire, people begin to manoeuvre for position.

Succession problems varied in intensity from firm to firm. Notoriously one of the biggest dangers to a firm's welfare, management succession may have its worst effects years before it actually becomes urgent. The problem appeared to be particularly acute in firms where their ageing heads had tended to stop carrying out the essential task of interpreting the general situation of the firm (including its future structure) even for themselves; it was no longer their problem, they seemed to feel, but that of the next generation. Suspending some of the major functions of direction and concentrating attention on the succession problem itself had the effect among the top management of shrinking their commitment to the working organization and enlarging their commitment to their own career prospects and to political groups through which they attempted to exert control over the succession choice, or, more indirectly, over the conditions under which candidates performed. A statement of the situation in one concern may serve to show these interconnexions.

Among all members of the top management there was an enormous respect and regard for the head of the concern, as a repository of wisdom and experience accumulated over very many years—almost the whole lifetime of the industry—and for his commercial flair and technical insight. 'In former days, A. K. was general manager, sales manager, chief engineer, commercial manager, chief accountant—the lot. Being the almost-genius that he was, and still is, he carried it.' At the same time, the question of succession was obviously pressing, and was still unresolved. This put an extra premium on good and close working relationships with him, beyond those necessary in any case with the executive head and arbiter of careers.

Immediately below the Managing Director were some half dozen Senior Managers. These formed the majority of a recently constituted Management Board, now declared to be a major planning and administrative body, acting under the Managing Director. They were also a majority in the Management Committee of the main works. General Managers of the other plants attended the Board; four

Divisional Managers attended the Committee. This dual arrangement had been intended to signify a general shift in the centre of gravity downwards and from a personal to a group responsibility for direction.

This shift was only partially accepted, or rather was disregarded operationally. While one informant said that a big change had occurred in that the Managing Director now had only six people reporting to him, whereas previously he had had thirty, another cited instances of his 'producing ideas' about redirecting effort or moving people some stages down, and of the Senior Managers responsible 'failing to persuade him' to adopt some other course. Again, while another said that the Managing Director had had to 'change over from the personal control to the delegated control', a junior member of staff who came in during the conversation was instructed to take an expense account in to the Managing Director and get him to sign it, since it was over the amount the Senior Manager was authorized to sign.

The four Divisional Managers were said to be 'responsible for' different aspects of their jobs to the six Senior Managers. But this relationship was hard to discern outside the protocols issued by the Managing Director to establish the new Management Board and Committee. The protocols existed as a paper constitution by which Senior Managers might claim 'line authority' over Divisional Managers. But the boundaries of functional authority had never been fixed or defined, and shifted from time to time. More important, they were liable to be overridden or ignored by the Managing Director, who by-passed them to deal directly with Divisional Managers, and by-passed Divisional Managers to intervene in the affairs of the Divisions. Caught between the arbitrary rule of the Managing Director and the pressure of the Divisions towards autonomy, some Senior Managers felt their positions to be dangerously marginal.

In practice, and this was probably cause as well as effect of these complications in the line structure, the functional network was overlaid by a fairly explicit political alignment. The Management Committee split politically into three groups: the older men approaching retirement themselves, but all still with several years' service ahead of them; the younger men, recruited since the war; and the new men, who had come in from other concerns. Each was led by a candidate—unofficial but widely recognized—for the managing directorship.

Because of the peculiarities of the distribution of functions among top management, and because also of the indeterminacy about the future leadership of the group, the structure of relationships at the top was rather more than usually complicated. The simplest—and one of the commonest—forms of top management organization is a series of single links between the Managing Director and each one of the members of top management; there is, of course, a constant and understandable tendency for both sides to develop this link at the expense of other links. It is simpler for both Managing Director and for senior manager —in the sense of 'knowing where you are'—to deal with one man than with several simultaneously, if either wants to get things done. 'These discussions (about the need for sales efforts as against engineering effort) go on within the Divisions and with A. K.—*not* in the Management Committee. If I can persuade A. K. that's good enough for me.'

The Managing Director was the central source of visible power ('Here the Managing Director is all-powerful') and by his Company role was the medium by which the Works influenced Company decisions and received policy leads; his approval, either implicit in a general and acknowledged association with the individual manager, or explicitly given to decisions or proposals, was the best possible validation of managerial action. On the other hand, a structure which allied each senior manager more firmly with the Managing Director than with his colleagues reduced the absolute and relative isolation characteristic of the top position.

This kind of set-up appeared to remain as the fundamental structure of top management relationships. 'And still today his is the hand on the tiller—there's no question about it. And if A.K.'s hand left the tiller here today, my personal belief is there would be utter chaos.' One of the weaknesses of such a structure, however, is that it may lead to competition for his approval and his attention; in so competing, and also in order to protect positions against the day when succession will have to be decided, the 'political' alliances had been formed. These represented a second structure, since much of the purpose for which the alliances were formed was necessarily to control access to the Managing Director, to strengthen and co-ordinate claims on him, and to attempt to exert control through him. Thirdly, there was the formal structure, with two overlapping committee groups, and a multiplicity of lines of responsibility. Each of these three structures was both a consequence of, and grounds for, the existence of the others.

The complications and pressures at work in this structure had consequences for the role of the Managing Director himself. Preoccupation with the choice of a successor had given increasing prominence to the single function of inspecting the performance and continually assessing and reassessing the relative merits of likely candidates. On the other hand, perpetually recurring involvement in the decisions and proposals of individual Senior Managers reduced his chances of assessing their merits in independent roles.

His task had been reduced to solving the succession problem and keeping going until it was solved. The evidence for this conclusion was provided by conversations both with the Managing Director and with senior staff. From these conversations it appeared that he saw the current overall task of the concern as a holding operation, the assumption being that the growth curves, interrupted in recent years, would shortly be resumed, and that business in the same technical and commercial terms as previously would again prove an entirely adequate field for expansion or survival. In pursuance of the holding policy, some steps were being taken to improve administration and financial control, but there was also much resistance to new proposals under these headings, largely because the general task and prospects of the concern appeared to remain unchanged.

The steadily increasing importance of the succession problem and preoccupation with the relative placings of candidates had its effect in another direction. Much of the consultation initiated by the Managing Director was thought to represent covert attempts to test people out: 'He asks me a question about something entirely in my sphere, and then goes to X and asks the same question.' The placings themselves had admittedly changed more than once in recent years, and not only for the very top position, but lower down. What he described as 'hopes for people' had changed. The consequences seemed to be that no one had any clear idea of the qualities regarded as appropriate for advancement to higher positions; more, people were beginning to suspect that these qualities changed from time to time.

DIRECTING THE ORGANIZATION

Over and above finding solutions, either in mechanistic or organic terms, for the complicated and manifold problems of the management system (which has to do with controlling the interpretive process) the Managing Director, alone or in consultation with his subordinates, has to see to it that the total usable resources represented by the working organization are properly deployed so as to meet the opportunities of its technical and commercial situation. The aim of top management, in this respect, was analogous to that, say, of a military high command: to create a manoeuvrable, aggressive force out of the resources at its disposal. Indeed, some managing directors discussed this aspect of their work in military terms.

This aspect of direction bore directly on the working organization when the concern was operating on more than one 'front'—i.e., was manufacturing products for significantly different kinds of customers. All concerns in the electronics industry which were studied did, in fact, operate in this way. The question then appeared of how the total force could be organized so as to make for most manoeuvrability (or 'flexibility') and aggressiveness (or 'initiative'). The question was posed with especial force in the larger concerns with factories, laboratories and offices dispersed in different buildings and often in geographically separate sites.

Two distinct principles of direction were apparent.

Central Control

The first was for tight central control with the object of enabling production, resources, managerial and design effort, and even the firm's objectives themselves, to be changed quickly as the situation demanded. In one of the larger companies not only did the board of a parent company dominate the boards of the subsidiaries, but in addition there existed a complex of committees, headed by a Group Management Committee and including a Group Finance Committee and a Management Committee for each subsidiary company. The main board had *ex-officio* membership of all these committees. Not surprisingly, the division of the Group into companies was said to be a fiction. Nevertheless, the operating companies served different sales organizations and were themselves distributed about the plants maintained, at some distance from each other, by the group.

The whole organization was directed in a very flexible manner. Given the point of view of the main board, it is inconceivable that any other arrangement could have been more flexible. Production lines were moved from one factory to another. Promising lines of development work, such as that for instruments, could be moved from an outlying establishment to the main plant, and 'given every opportunity to expand'. Attractive solutions could be found for other managerial problems along the same lines: the executive who did not 'quite fit in, at one place, could be recommended to another as a 'jolly good chap' technically' and well worth a chance, and even worth promoting—so as to make it look well—at another plant.

Yet such escape-hatches were not the best way out of the difficulties themselves, and resort to them had a depressive effect. In one instance, a succession of solutions to various problems of seniority and incompatibility by transfer resulted in the paralysing juxtaposition of two young and ambitious works managers in the same plant and an artificial division of 'technical supervision' and 'production administration' between them. One establishment saw itself as 'working for three sales organizations'. Not only had production lines and people been moved to and from this establishment, but a prospering development section had gone to headquarters. This gave rise to some feeling about the establishment's being treated as an old-style colonial dependency, impoverished by having to yield its most valuable assets to the 'mother country', thereafter having to be treated, as it had recently been, as a 'problem area', to be helped out with aid programmes such as the transfer of additional production. This feeling was exacerbated by the necessity of referring decisions of any significance to headquarters, and by the knowledge that the management was cut off from the political centre of things. There also tended to be sharp resentment against 'visiting inspectors from the board', who appeared frequently on the premises in order to 'maintain liaison'.

Quite apart from the restriction on enterprise that these circumstances engendered among the outlying plants, senior managers and development engineers in them felt frustrated and suspicious and away from the centre of things. Since the main route to the top, or even adequate control within their own sector, seemed possible only if they were at the main plant and in the political swim there, they tended to look for opportunities to demonstrate the disadvantages of the present location of themselves and their sections and the benefits that would result from their removal to headquarters. The repercussive effects of such moves long before and after they occurred were acknowledged to be unsettling and costly.

The second kind of directive principle was founded on the belief that it is impossible to direct all the concern's activities from a central position and that it was better to create sub-enterprises capable of discharging all, or as many as possible of, the managerial functions in respect of a specific range of products made by the firm, reserving some functions, and sometimes production, as common services. This latter type was further subdivided, some companies operating with product divisions, others setting up small enterprises, often with the status of a separate company, which were given a great measure of freedom.

Sub-enterprises (i)—Product Divisions

Product divisions seemed to be most favoured in situations which had appeared to a new managing director to call for speedy reform. Reorganization begins at the sales end, which has the effect of emphasizing the commercial purposes of the firm. The next move is to realign development laboratories along similar lines. It is here that difficulties begin. The proposed solution in one company has been stated in these terms:

'On the Ministry side the pattern is a compromise between self-contained projects, which appeal to the Ministry for security and continuity, and specialist services which could be more economic. In practice centralized services are used where the knowledge or the laboratory equipment are peculiarly specialized, as in the microwave and the climatic laboratories.

'As each commercial project has arisen, it has been based on the exploitation of a technique, and hence has started with a development team conversant with that technique. But in the design stage the technique may be of much less importance than familiarity with the environment, both pysical and mental, in which the equipment will be required to operate. Such familiarity is special to the product and to the type of customer, and such development as it involves, and the initiation of research work which it implies, should be centred on the particular product group.

'From time to time, new products will be initiated for which no product division has been formed. In this case development and design may have to be done in a centralized Advanced Development group, but as this group may lack the specialized knowledge of the customer and his environment which is the equipment of the product design engineer, steps will be taken as soon as possible to build up a definitive team, perhaps nursed initially within the central group. A central group will also be required to give specialized or temporary assistance to product designers, and co-ordinate development of techniques which are common to two or more product divisions, as well as acting as liaison between design and research on new techniques which could solve design problems or assist in finding a market.'

The indeterminacy of direction in this account is typical. Underneath the practical appreciation that every development and design project is different from the last and may need different kinds of managerial and engineering attention, there is the fact that until recently the concern was dominated by a large and powerful development laboratory, organized mainly in terms of technique. The commerical situation had changed: product divisions had appeared in the sales force and were now beginning to create small dependencies for themselves at the 'design end of the laboratories' where 'technique may be of much less importance than familiarity with ... the product and type of customer'.

There is always, in fact, an ongoing organization which direction is struggling to adapt to the tasks of the present and of the future, as they make themselves known. But when product divisions are created as a response to the increased importance of the sales task; when this importance gains recognition in an increase of the size and a rise in the status of the sales force; when directorships are offered to salesmen outside the firm who are credited with the measure of forcefulness and experience needed to enforce the change in direction internally and externally, the whole issue of how the firm is to confront its new circumstances is transposed into political and personal struggles for power. It was in the firms which had reacted to the creation of product divisions in order to implement a direction change that the political conflict between sales and development was most apparent.

Sub-enterprises (ii)—Dependent Companies

A few concerns have followed a principle of creating new companies, or fairly autonomous sub-enterprises, to deal with isolable sections of the total task. Such a section may be engaged in assembling and selling a range of products (e.g., instruments, transistors); or developing and marketing new products based on new technical applications (e.g., in the data-processing and data-handling field); or manufacturing metal frames and components. This implication of the senior management in the fortunes of actual products is a logical extension of the conventional grounds of business enterprise. It certainly makes for close identification of staff with the activities—sales, design, production, financial control—of all sections of the concern. Yet there remain certain essential conditions for the success of this kind of strategic disposal of resources, conditions which once again reveal the imperative need, in directing a firm, to relate the internal system to the external tasks. The different experience of two multiple companies provides an illustration of this elementary principle.

The first—Alpha Company—was split into five or six companies, mainly for the reasons we have already enumerated, but partly because it was a declared belief that a size of five to seven hundred people is as much as one can hope to reach and yet provide really favourable conditions for personal satisfaction while having regard to the technical requirements of an electronics enterprise.

Yet allowing for some difference in the circumstances of the two concerns, similar directional arrangements in another concern, —'Beta Company'— were said to have proved utterly unsuccessful. Until recently, Beta Company had been organized into separate companies, each based on an operational phase: research, development, production, and sales. Each firm had been given virtual autonomy, and was expected to operate as an independent commercial firm and thus to show a profit on all transactions, whether they were conducted for, or in concert with, other members of the Company or not.

This policy was later universally condemned in the Company. One or two thought it 'a wonder that the firm had survived this period'.

'The general arrangement was that sales people would ask for something to be designed, research people would try to make a profit out of their end of it, the design people would try to make a profit out of doing their bit, the factory would try to make a profit out of producing it, and if at the end of all that it didn't cost about twice what any customer would pay, the salesman was left to sell it —and make his profit. . . . In fact what happened was that the sales people ignored electronics and concentrated on electrical products, radio and television, and eventually succeeded in getting hold of their own factory and their own design section and operated successfully. The research people went out and got Government contracts which would pay costs and a percentage without the inconvenience of some little man around the corner grumbling . . . and the Development Company did the same with Government development contracts, and under extremely energetic management expanded itself out of all proportion to the other bits. The Production Company was left with no option but to take what Government contracts it could as well, staggering slightly because the ones it tended to get were those that didn't result from the development contracts taken on by the Development Company. The development contracts were taken on in order to show the biggest overall profit on development, not necessarily to show the biggest profit for the Group. We got a reputation for taking on the sticky, awkward jobs that all the more tightly organized companies had turned down—because you were going to spend an enormous amount of money on cost plus 7 per cent development work and then be asked to make half-a-dozen in the end.

'All this led to extreme suspicion and bad feeling between the companies . . . and did us no good with the customers, because you would find the chaps in the Research Company and the chaps from the Development Company would go along and have a slanging match with each other. And the production people too: it wasn't their fault if production was late or things were badly made, or that they cost too much, it was those stupid so-and-so's who had done the design—that sort of thing. . . . This actually happened. I've heard it happen.'

The success ascribed in the one concern, Alpha, to a directional policy of creating subsidiary companies is in striking contrast with the universally agreed failure of a similar policy in the other firm, Beta Company. But in a sense, the contrast provides the clue to the distinction we have drawn between organization and direction.

There were, of course, differences in any case. The major firm of Alpha Company was the parent company itself; it included no design or production activities, but provided services; accounting, purchasing, personnel, commercial, sales: and facilities for carrying out market, organizational, and other studies. It also acted as a traffic centre for communication, as did the main board for decisions which might have any repercussions on other companies and the group.

But there were also similarities beyond that of structure. In Alpha Company, too, rivalries existed between the companies which could not altogether be classed as healthy competition: there was resentment among older groups about the allocation of resources to newer enterprises which had not proved themselves; members of one company might feel underprivileged compared to those of another which was better housed and better equipped; some hostility existed between the subsidiaries and the parent company over the part played by the central sales organization in their individual affairs.

The point of variation, for us, lies in the difference between organization and direction. In Beta, division into sub-enterprises cut the organization into three, while the commercial strategy remained identical for all. Almost the whole of their work was done for two markets: government and domestic radio: which all three shared. Not only, therefore, was the directional policy pointless; it seriously disrupted the organization, which had to bridge profound gaps in the interpretive system. In the other, successful case, Alpha, while there was one gap in the interpretive system—between the end-product companies and the production companies—the fact that none of them were rivals in the same market justified the directional policy of division.

There is also the different size and ages of the companies involved. The companies of Alpha Company were all small, young, and growing very rapidly. For each one of them its situation at the end of a year was utterly different from that at the beginning. Whether or not it would double in size, merely arrive, or even fail was always doubtful, although plans were normally based on expected growth. The companies in Company Beta Group were parts of a comparatively old-established concern, and the opportunities and direction of growth had been, in government work, largely determined by circumstances outside their control.

According to the evidence supplied by these studies, and the interpretation of it given here, both the organization of the internal interpretive system and the direction of the commercial, technical, and productive capacities of that system in the interests of the firm are conditional for their success on an appreciation of the rate of change affecting the tasks of the concern: i.e., of the number and importance of new technical and market circumstances confronting the firm from day to day. *As the rate of change increases in the technical field, so does the number of occasions which demand quick and effective interpretation between people working in different parts of the system. As the rate of change increases in the market field, so does the need to multiply the points of contact between the concern and the markets it wishes to explore and develop.* This last consideration weighs very heavily in deciding the success or failure of directional policy in technically progressive firms.

CHAPTER 11

The Shaping of Work Relationships

Most managing directors have risen from the lower managerial ranks. All but three of the heads of the sixteen concerns in these studies had reached their position after very long service with their company. They were, to a very large extent, what the system had made them; they had arrived by accepting its values, promoting its interests, cultivating and grooming those capacities and traits which made for success in it and extinguishing or concealing those which did not. This observation, which is trite enough, is not intended as any invidious comment on either managing directors or the system. Exactly the same thing could be said of professors, bishops, and permanent under-secretaries. Rather, we are using it to underline the limitations of their role. Even revolutionary changes have to be in the direction of one of a handful of familiar variants of either one or the other kind of management system. More important is the manager's capacity to retain, and perhaps develop, the self-conscious perceptiveness and single-minded opportunism by which he succeeded and which he must now finally and unreservedly apply to the external situation; failing in this he can divert, frustrate, or waste the energies of his subordinates. More important still is the aptitude he possesses and the skills he has learned for instigating commitments to the working organization and its tasks.

This second aspect of direction occupies these last two chapters.

In the brief account given in the Introduction (pp. 12–14) of the methods of gathering the information we thought relevant to this enquiry, we observed that the starting-point of our attempt to explain the situations we encountered usually lay in the differences between the descriptions of the same jobs, or of the functional relationships between people and departments, which would often be given by the individuals most directly concerned. We became very accustomed to being put right by managing directors in our accounts of functions and responsibilities, accounts which were in fact supplied to us by people whom we asked to describe their own jobs.

The question which we tried to formulate in these cases was not 'Which of these two versions is the right one?'. The 'right' version, only too obviously, is that which conforms with the view of the head of the concern. Instead, the question we thought relevant was 'How is it that these two versions of the same set of circumstances have arisen in the minds of people who have to co-operate with each other in the very circumstances they view so differently?'.

THE INDIVIDUAL AND THE MANAGEMENT SYSTEM

The crucial social fact about the working organization is that while, for others, each individual is one bit of the system with which he has to deal—one sink or source of material or information on which they themselves work—he, for his part, is dealing with each of them as bits of his personal system. (This perceptual fact has, indeed, been expressed graphically in a series of organization charts drawn by junior members of a factory, each of whom showed his own department—transport, accounts, machine shops, assembly department, production and engineering, etc.—as the centre of the organization. Other parts were rendered as more or less peripheral, according to the amount of traffic between them and the 'central' department.)*

Each individual, in effect, has a *whole* job to do, and to complete his tasks and solve his problems involves exploiting other people—acting through the mediation of other people's jobs. The exploitation of others in the fulfilment of one's own job is most clearly realized in the behaviour of persons towards subordinates: 'that is what subordinates are there for'.

In stable conditions, this dependent form of subordination appears to be acceptable and effective. In changing conditions it is not acceptable and hinders effectiveness. So far as the members of organic working organizations operated effectively, the exigencies of their roles were constantly involving them in actions through which they openly or tacitly rejected subordination.†

But the rejection of subordination is complementary to greater

---

* Personal communication from Mr. Iain Hird.
† For the expression of these observations in numerical terms, see T. Burns.[87]

commitment to the working organization. Indeed, it follows from the greater immersion of self in one's job; one can no longer practise the self-detachment by which ordinary people accommodate themselves to the routines of submission. When the individual is involved in the bigger, more active communication network required for faster technical and commercial change, he is more fully implicated as a person, more committed, more involved.

The growth in importance of the occupational self implies a closer identification of the whole person with work; the centre of gravity shifts significantly away from family and the outside world and towards his working life. The notion of subordination itself becomes less tolerable and is often rejected, or set aside by collusion between the members of the organization. Yet, at the same time, his closer incorporation in the system involves a loss of personal autonomy. Not only is the life he leads and its emotional and intellectual content more contingent on his participation in the working organization; he is drawn more frequently and more closely into personal relationships with the other members of the organization. He is no longer 'left to get on with the job himself'.

As the communication system implicating the individual widens, and he becomes a more active element of the larger system needed for work to be accomplished at all, his share in the outcome is diminished. In this, management reflects the institutions of technical development itself. The individual technologist is backed by an extensive communication network through which he keeps informed of increases in technical information and scientific knowledge; and he becomes less of an inventor. Again, because of the very indeterminacy of the information which might be necessary to complete tasks, more demands are made on each person; he is required to devote more and more of his time and mental capacities to functional activities which are less and less his own sphere.

The shift from mechanistic to organic procedures, therefore, makes considerable demands on individual members of an organization. In general terms, they are required to surrender the safe determinacy of a contractual relationship with the firm for one in which their obligations are far less limited, to replace a view of the firm as an impersonal, immutable boss by one which regards it as something kept in being by the sustained creative activity of themselves and other members, to cease being 'nine-to-fivers' and turn 'professionals'.

TECHNICAL CHANGE AND THE INDIVIDUAL'S JOB

In one instance this process of adaptation was visibly under way at every level. The concern had been established two or three years earlier to make miniature electrical components for a parent company. It had settled down fairly quickly. Design was fairly stable, programmes were composed in accordance with expectations of steady expansion; although processes and assembly were fairly intricate, the firm attracted workers who acquired the necessary skills fairly quickly.

Just at the time when the phase of first growth was ending, the design of the component underwent a revolutionary change, the outcome of many years' work in the parent company. The company handed over the entire work of bringing the new design into production to the branch establishment.

The first overt change was the recruitment of three or four industrial scientists with appropriate qualifications and some special experience. To begin with, the new group operated as a separate development laboratory, a service department called in to instruct in the use of a new manufacturing equipment, to deal with problems arising in the course of production or out of test or inspection results, to institute modifications in design or production methods. There quickly ensued a political division, the 'old hands' among management resenting the intrusion of the new group between themselves and the factory manager. As one or two of them later said, he was spending 'far too much of his time with the lab people'. Meanwhile, production difficulties multiplied, and the 'lab people' were often conspicuously unsuccessful or slow in getting them remedied. Lastly, the distinctions of status and demeanour assumed by the 'lab people' made themselves felt; their 'superior' attitude to the production side and their lateness in coming into the factory were especially resented.

Eventually, the factory manager decided to try to settle these differences and to work out an effective mode of operation by direct methods. He called a meeting of both sides, at which he quoted complaints made by each against the other. This succeeded in making the breach public and it was widened by some additional criticisms made at the meeting. Nevertheless, each side was confronted with the knowledge that the others had counter-charges, or even the semblance of rational defences. This offered sufficient grounds for the factory manager to attempt a revision of the working organization. He had come to the conclusion that it was impracticable to maintain the old mechanistic system, with each person in the production hierarchy, down to assembly hands, discharging a clearly defined, pre-planned job. With the co-operation of the local Technical College as well as of the firm's industrial scientists he instituted training courses, not only in the technical principles of the new device but in management methods. Secondly, the division of the production side into small departments, each responsible for a single part of the whole process, was dropped, and a large part of the staff of departmental supervisors, charge hands, and operatives was re-formed into so-called 'development groups', each with the task of solving production problems. This meant that any individual member of the production staff might not only move from one stage in the process to another, but might also move from routine production on his or her own to participation, under a production manager or laboratory engineer, in a problem-solving or development team.

In brief, what the factory manager was attempting was the conversion of a small manufacturing plant into a fairly large development establishment. In the process, most jobs shed their specific, contractual nature, and took on an ill-defined, elastic range of activities with people having to co-operate to find answers to questions which had previously been posted upwards; similarly, communications based on a vertical flow of information up and instructions and decisions down gave way to a lateral system in which information was exchanged between equals and members of teams who were equals for the time being.*

Resistance to Change as a Defence of the Self

As an acceleration of the rate of change requires a change in the system of management, so does it force the individual into a new relationship with his work.

In many firms which were launched into electronics development, their individual members refused to accept the new relationship, with all that it implied. This refusal was implicit rather than explicit. It in-

* One may compare this brief account of a shift from a mechanistic to an organic system which affected all levels in the hierarchy of the working organization with the detailed analysis of similar systemic change in a work group of operatives in A. K. Rice's study of the change-over from non-automatic to automatic looms in a large Indian textile mill.[88]

volved managing directors no less than foremen. It could, and usually did, go with approval or support or even sponsorship for the new policy. It could, and did, involve the same individuals in harassing, even bitter, anxieties, animosities, and frustrations.

All this seems akin to a pathological adjustment of the individual—and, eventually, of the concern—to conditions of change. If one adopts the view that an increased rate of technical progress is healthy, or necessary, or desirable, then resistance is indeed pathological. But, equally, to resist adapting an organization and work roles to the demands of rapid change is a measure of self-defence, is a 'natural' reaction. When, for example, a need for more effective and widespread communication is converted into a demand for specialist groups of interpreters and intermediaries (pp. 168-71), the actual and manifest changes which occur are seen as threats to the existing structure of power and status.

Other individuals and groups are not only capable of resisting, but may invoke departmental loyalty, self-interest in one's own career, and perhaps 'the good of the firm' as motives for resistance. It is not political action as such, therefore, but creating institutions within the concern or altering the balance of institutions to embody emergent needs which is the strategic key to the resistances.

In putting this construction on events in the concerns studied, a clear distinction must be kept between the manifest purpose of people's conduct—what they have in mind, and its latent function—what their conduct is, perhaps unconsciously, designed to achieve from the point of view of their total career, occupational, 'social', and domestic.

ORGANIZATION FOR RESISTANCE TO CHANGE

The causal sequence which this interpretation imputes to the course of events and the reaction to them observed in a score of industrial concerns is revealed better by a detailed examination of one example than by attempts to present summary revisions of the material already used in previous chapters. The example we have chosen is not representative. It is the concern which became involved in the biggest change in the rate of technical progress which we encountered. In a variety of ways, the management—and particularly the top managing groups, including the managing director—sought to maintain a particular variant of mechanistic organization, despite the great organizational and personal difficulties to which it gave rise in the new situation.

The concern was an old engineering firm which had revived during the war and had since grown larger by building up a sizeable business in fabrications, although the market for its main products was declining. During the war years, and immediately after, the management system had remained virtually unchanged. Much had depended on the specialized engineering knowledge and energy of the Works Manager, who had been recruited early in the war, and who succeeded later to the post of Managing Director, and he had apparently adapted his procedures to those he found in the workshops, which were aspects of an unusually stable situation.

The firm had manufactured virtually the same range of products for two or three generations; design improvements had been evolutionary and slow, and related to general developments in craftsmanship, materials, and machine tools rather than to technical progress within the firm. The fabrication work which the war had brought in depended more on recent advances in welding technique, but this information had been supplied by the head of the concern when he had served as works manager. Also, it had been 'fed in' at the point of the individual performance of craftsmen, not through design, so far as the firm was concerned. Designs were supplied by clients.

Until very recently the firm had relied heavily on traditional forms of supervision; a Production Controller appointed in 1953 said:

'I should say, looking at the history of the firm as I know it, that the works managers that they had here in the past had been relying upon the foremen for many duties which we feel today we should think for the foremen before all the responsibility is placed on their shoulders. Before I came the foremen were asked to estimate for work. If the order came in, they were then asked to put it through the shops and to make sure that it didn't cost more than what they had stated it would cost. It seemed to me from examples which we obtained that the foreman, naturally, if a job came in, would be conservative in his estimate. He wouldn't be too close for fear that if the order came in he felt he must have a little in hand. And we proved, too, that—Look, the foremen were very honest men. They still are. But if it was found that the job was costing a little too much, then labour might be switched from one job to another to make one balance. They had felt, from their point of view, it didn't really matter. Provided they had a hundred pounds, what did it matter if it was fifty pounds here and fifty there.'

Interviewer: 'They were operating in fact as though they were sub-contractors?'

Mr. X: 'Up to a point, yes. . . . But I think in most old-established works the foreman was left to do these things, and I've worked where the situation still exists that way where the manager would say "Look, Jack, we've got fifty pounds to do this thing. Just do your best", and it's up to him how he does it.'

Complicated as the processes and assemblies were, long-established routines which were now part of the skills of local workpeople allowed progressing and other forms of control to be built into the foreman's and departmental manager's job; abnormalities and departures from the routines of production would be immediately obvious. During the war, when the new fabrication work was added, there was little room for change in this system, even had a need for it been seen. Nevertheless, during the war and for five or six years after, the work of repair and fabrication which engaged most of the firm's efforts necessitated extreme flexibility in production control. Prices depended on speed; steel and other shortages made 'cannibalizing' usual practice. Some orders had to jump production queues, and since this meant that other work would have to be postponed and some orders superseded, the elasticity would have to be fairly covert. Jobs had to be steered through a maze of different prices, exiguous stocks, and all manner of hold-ups, without any governing formula and without informing customers continually seeking news of the progress their contracts were making. This was done by the constant personal supervision of every production and office department by the Works Manager.

The system of management, therefore, was a special type of mechanistic structure. What each person in the management hierarchy had to do was defined by the head of the firm (as the Works Manager in practice was, so far as inside affairs were concerned, even before he succeeded to the position of Managing Director). Moreover, even within their functionally specified activities they depended on his daily instructions and decisions. His managerial role, so far as the works was concerned, was acted out in the constant presence of subordinates; all foremen and some skilled hands were in daily touch with him and 'took their orders' directly, and daily, from him by word of mouth. For four years, indeed, he lived in a house within the works itself.

For each individual, therefore, there was an extreme simplicity about his place in the management structure. Outside his own technical knowledge and experience, there was only one source of essential information; if he kept in touch with that source, then he could manage on his own, or virtually so. 'Managing on his own' meant that he could make decisions and choices within what he felt to be his field of competence; where he was doubtful about the wisdom or desirability or feasibility of any action, he could consult the head of the firm and be provided with information or armed with an instruction. Conversely, the same relationship was there for use to put the system to work so as to meet requirements arising in his own field of competence which he could not meet from his own resources.

The three fundamental elements of the procedure, from the individual's point of view were:

(*i*) The continuous command exercised by the head of the firm meant not only that he could be treated as a perpetual reference manual of limits and constants and of codes of practice, but that one could define for oneself the safe limit of action and decision, 'posting' the rest upwards.

(*ii*) While the system implied that all managers and supervisors were the instruments of the head of the firm, he was equally theirs in the conduct of their jobs.

(*iii*) To deal with any management problem only two people, in principle, were needed; the head of the firm and the subordinate to whom the problem presented itself, or who was concerned in its solution. The knowledge that in any one situation the individual only had one person to deal with within the management hierarchy— and thus one person whose reactions and expectations were important—left him a very considerable degree of command over his personal situation.

This simplified statement may be misleadingly unrealistic if two fairly obvious qualifications are not added. First, the normal 'field of competence' we have referred to as the individual managers' concern included some traffic of information between themselves and other managers; but this would also be conducted in the context of pervasive control by the head of the firm, and in the knowledge that one was only 'on the safe side' if he were consulted or informed. Secondly, the manifold pair relationships between him and subordinates which constituted the basic management system do not have to be regarded as necessarily sympathetic or amicable—although there is no reason to believe they were not; the crucial circumstance was that in dealing with each other, it was possible to arrive at a working consensus concerning mutual expectations and the extent of each one's contribution, especially when subordination underlay the relationship. Further, any hesitations, errors, or failures could be translated immediately into projections concerning the hostility of the other; in a more comprehensive system of work relationships, the individual would find himself confronted with a majority judgement and be driven to dissociate himself from his milieu in order to dispose of the combined judgement of others on his personal inadequacy.*

Shortages of materials, the release of an enormous demand for capital goods, and the reorganization of large sectors of industry because of peace-time reconversion and nationalization, all combined to maintain conditions in which the wartime methods of management survived unchanged for several years.

Not until after 1950 did changes in the external environment begin to affect the concern's internal order. Government orders for fabrications fell off, and were replaced by smaller orders with varied specifications and requiring individual planning. Standards of inspection became more stringent. Profitable orders for quick repairs to the firm's staple product dwindled, and production contracts were now drawn on prices determined by large-scale producers. The quest for new lines of manufacture and the growth in the range of products and markets detached the head of the firm for days and weeks at a time from the close relationships with executives which had obtained previously. New products were acquired, two subsidiary companies formed. At the beginning of the 1950s successive aspects of the firm's organization were suddenly transformed. Development laboratories were built, and two separate teams of industrial scientists appeared. The office and sales organization grew, stores and purchasing became much more complicated, demands for unfamiliar types of production work began to arrive in the shops.

Strains quickly made themselves apparent, first in the production

* For a further elaboration of the qualitative differences between the relationships of pairs of people and those of larger groups see G. Simmel.[89] See also, for an account of a successful firm in which a management organization similar to that described here was founded on admittedly hostile or harshly authoritarian pair relationships between the head of the concern and executives, C. Argyris.[2] For social contrivances which enable individuals to preserve their self-images in the face of failure, see T. Burns.[1]

shops, inevitably (see p. 163), and in the lower ranks of management. A new Works Manager insisted early on that he would have to institute more stringent inspection methods as a counter to the higher standards now imposed by the main customer for their staple product. Talking this over later, the Managing Director said he hadn't liked the idea to begin with but was finally persuaded and gave way.

'It's true to say', the Managing Director went on, 'that there's been more trouble in those shops, more complaints and so on, about products coming from those shops where they had inspectors, than we had ever had before. Before then the supervisors and the foremen used to do the inspection. For over fifty years the foreman has been completely responsible for all material sent out of his shop on to the next one. It's true that two years ago standards were tightened up, but the same sort of thing is still possible. What had been an eighth of an inch tolerance, for example, is now a sixty-fourth, but there isn't all that extra burden. When the inspectors were first brought in foremen and supervisors still reckoned that they were responsible for stuff going out in the right shape—as in fact they were, of course. So in time you had inspectors not bothering to look too carefully because they thought the supervisors would do it, and the supervisor not bothering because he thought the inspector had, and the foreman didn't bother because, after all, two people had looked at the job before he had.' In consequence, he had found one Monday morning that a job had been loaded on to a lorry and sent off which apparently had never been inspected by anybody. Because of this incident, he went on, his Works Manager had been absent, spending the day with the customer, when a walk-out had occurred in one of the main shops.

Changes in the firm's relationships with customers, the new demands for a flexible production programme which could cope with variations in the nature and quantity of manufactures, also gave rise to the introduction of a production controller.

Production control also met with difficulties. The practice was still maintained of picking up some orders for fabrications on the strength of a promise of early delivery. The firm's Commercial Manager said, 'There's a technique about getting orders for fabrications. We'd get to know through our connexions of a firm which was pushed pretty hard to get some fabricating work done which they couldn't do themselves. Shipyards who were afraid they couldn't meet launching dates were often in that particular trouble. I will get in touch with their Works

Manager, because he's the chap the pressure falls on, and I'll probably go down and see him. I'll talk it over with him and of course with as many other people there as I could, and I'll get the idea that the job's there for the taking—we might pick up the idea that the firm was in a bad fix and they'd have to get this fabrication done, come what may. Well, of course, the policy then is to impress them with the fact that they are being offered a favour, and that if we did take the job on it would disorganize our own production schedules and so on and so forth. You see, in the upshot, next time we went looking for an order from this firm, we wouldn't have to go down cap in hand but we'd expect to get something in return for having helped out before.'

We asked how it was possible to take on such jobs when they would in fact disorganize production. He said, 'That's the technique. You have to impress them that we're going to have to disorganize our own production.' And in answer to further questions, 'We talk this over, the head of the firm and I, and then if we decide we can take it we call in the Works Manager and tell him. I say I've given my word that we are going to do this job, and he has to get on with it.' This sort of decision by the head of the firm and the Sales Manager was apparently taken on the basis of 'looking at the state of the order book and delivery dates, and of course the head of the firm gets around the works every morning and knows what's going on.'

Planning and progressing was applied systematically only to the staple product and to fabrications. When the concern began to manufacture new products, the planning and progressing of any parts made in the main shops was the work of a 'co-ordinating manager', who worked on the other side of 'a dividing line between what is the factory's work and what is new work so far as estimating, production control, progress, is concerned', as the 'co-ordinator' put it. Works orders and job numbers for such work were issued by the Production Controller himself, although the co-ordinator had his 'own progress system. But I get together with the Production Controller and discuss the capacity of the works, so that he knows where to serve out drawings and jobs and so on.' On the other hand, the co-ordinator was under constant pressure to gain priority for certain jobs, and was also often obliged to have parts remade because manufacturing needs ran ahead of the production of drawings. As one shop superintendent said: 'There's always some trouble on this new stuff. As an example, the only thing of theirs here at the moment, there's a large casing, the top one of a special job most of their work is concerned with at present: it had to have some extra castings welded on at varying angles. No detailed drawings were given, so they were all welded at a similar angle, then they had to be removed, different cuts in the main case had to be made, and new castings made and fitted. . . . Some of this stuff needs 'knacky' welding, and they're going to be in a bit of a pickle when my only knacky welder leaves—he's going in a few weeks.' Experiences of this kind, and the tendency to regard jobs put in from the new groups as 'extras', meant that priorities gained for them by the co-ordinator would frequently be set aside.

Throughout the working organization, then, the system of operation which had formerly obtained still persisted. But even so, it persisted in piecemeal, unco-ordinated fashion. Inspection, typically a 'lateral' relationship, failed to work efficiently; the Managing Director's dislike of the separation of inspection from supervision possibly contributed to this failure. The Managing Director's authority could be invoked to disrupt production programmes; even more significantly, the same arbitrary and disruptive decisions could now be made with impunity lower down. The only remedy for the victims of these decisions was to apply to the Managing Director—a procedure which prevented the growth of a more effective or economical system of working relationships.

So far, the account of responses to the changed circumstances of the firm has dealt entirely with the vicissitudes experienced by the older part of the organization and the measures taken to overcome them; the new accretions were involved in one or two of the new systems— production control and budgetary control, for example—but only incidentally. The other aspect of transition, the incorporation of the new technically progressive parts of the working organization, was much more fluid, and less easy to resolve into specific problems susceptible of known solutions.

As we have said, the characteristic feature of the earlier management system was the series of pair relationships, one end of which was always the head of the firm; each relationship existed in its simplest terms as a manipulation by each of the other in the interests of his own attainment of the ends, as he saw them, of his functional activities. In terms of control over the circumstances determining performance in his job, therefore, each subordinate was very heavily dependent on the head of the firm, a dependence which was magnified by his being the source of most of the information which affected the conduct of their jobs.*

The technical and other managerial posts required by the new developments, as they occurred through time, were fitted in to this system of pair relationships radiating from the head of the firm. Several innovations requiring development laboratory work of a different technical nature were seriously considered as ventures by the firm, but eventually it embarked upon three, each of which was technically separate, with different development laboratories, separate laboratory groups, and separate technical managers. There were also different fitting shops and drawing offices and, to some extent, separate organizations for sales, production control, estimating.

Each person appointed to these various functional positions, or to the head of each functionally distinct group, was individually dependent for the conduct of his work on the head of the firm. Changes in the old management structure had involved a number of changes in personnel, and, some three years after the firm had begun investing in innovations, all leading executives—the group of eight who lunched together in the top dining room—except the Managing Director, had been recruited to the firm from outside within the previous two years. They were uniformly dependent on him at the outset for information, guidance, and decision-making; there was no one else who might, when they first came, have acted as an alternative source of information. This training relationship strongly reinforced their functional dependence on him.

In interview, every one of the senior executives, and many others, dwelt upon the impossibility of obtaining a working relationship with other executives at the same level. They would also, at the same time, speak of the number of people with direct access to the head of the firm. One of the industrial scientists made the point by comparing what had happened in the larger concern in which he had previously been employed.

* Dependence is to some extent the reciprocal of the instrumental character of the relationship: each person had to make use of the other in order to get personal tasks accomplished, and to this extent each was also dependent on the other. Nevertheless, since the superior was obliged, or was entitled, to restrict the information he imparted, while the subordinate was not (whatever he in fact did) the weight of dependence was unevenly distributed. There was also, of course, the subordinate's dependence on the superior's approval for keeping or bettering his position, a dependence which is the foundation for the functional relationship we are considering here.

'In X's, lateral communications were reasonably easy. You could get things done without going to the head of the factory or even the heads of departments. In fact the simplest thing to do was to go to somebody at your own level. It was easy to get a routine going. . . . You could arrange with somebody at your own level for a procedure to be followed whenever the sort of occasion arose that you'd agreed upon. All that is impossible here. There are no routines. Lateral communications just don't exist. There's only one sort of routine. That's the head of the firm, and it very seldom works. . . . He probably has thirty or forty people reporting to him and it's obviously impossible. You just can't get at him.'

Another senior manager said, 'Anybody can knock on the chief's door and get in. This is very much disapproved of. You get senior people not knowing what's going on. I insist on my own people telling me whenever he goes to them and tells them to do anything.' A third member of the senior group thought that the continued practice of the head of the firm's touring the works every morning 'complicated things'. He thought it gave 'an easy way out for people doing things differently from the orders given them by their immediate superior. They could always say it was the boss who told them, and of course, there's no come-back.'

The maintenance of the forms of direct supervision, by which a large number of subordinates were separately dependent on the head of the firm for definition of their tasks, methods, and field of control, inhibited the growth of any other kind of organization adapted to the needs of the new situation. In the part of the firm concerned with one of the new products, executive responsibility seemed to be dispersed uncertainly among five people.

(*i*) There was first the chief technical person, an industrial scientist who said he had come in to 'create' an organization which could develop and make the new product.

(*ii*) The second man involved was an executive, recently detached from production control in the factory at large, and now, he said, 'responsible for all production of this particular new product'. 'I'm responsible', he went on to say, 'to the head of the firm. . . . Oh yes, formally it's Dr—(the industrial scientist) but of course he's a scientific chap. He hasn't got much to do with production and he's pretty remote. He thinks up an idea on his own, takes it away and dresses it up on his own in the lab. I attend to getting the parts made up in the works or sub-contracted out. . . . Drawings also come to me. The Works Production Controller and I get together once or twice a week and discuss the capacity of the works, so that he knows where to send drawings and so on.'

In the shop where the product was assembled 'there is a manager chappie, whom I see quite a lot of'. But for the most part, he employed the pattern of working relationships customary throughout the concern. 'I like going down and talking jobs over with the people who are actually doing them. . . . Oh no, I make no bones about going straight to the men on the bench.' Above himself, 'I deal direct with the Managing Director'. This direct relationship was insisted upon later in the interview, and he made it quite clear that he has no 'responsibility towards' either the Works Manager or the industrial scientist who had been the founder, and was the nominal head, of his own section of the firm.

(*iii*) The 'manager chappie' in the assembly shop on the other hand, said he also had direct contact of a similar kind with the head of the firm, and moreover, had some grounds for claiming recognition as the 'head of the production side' for the new product, since he had been the principal negotiator with the accountants in the inauguration of budget control. So far as this kind of administrative responsibility went, he was also controller of ancillary laboratories, which were otherwise the direct concern of the industrial scientist.

(*iv*) Another executive, described by others as a sales representative, said that this side of the firm was really run by a committee, consisting of the head of the firm, the industrial scientist, the Company Secretary, the Commercial Manager, and himself. He said he attended not only to the commercial side of sales but to the 'administration of the process'. Later, in an attempt to make clear his relative position, he described his 'responsibilities' as 'divided between the Managing Director on the technical (*sic*) side, concerning individual sales, an outside Director with some working interest in sales concerning sales policy, and the Commercial Manager on the financial side of sales'. In reply to a question about the possible difficulty of knowing whom to consult when a difficulty cropped up, he said, 'Really, when you consult anybody you consult everybody. My great liking is for vertical organization down the direct line. . . . You can't really get a direct line in the place, but I don't find communication over-elaborate. I get my mail from the head of the firm, and all my outgoing mail has to be signed by the Commercial Manager. There is no trouble about this. Anything which might be queried by him I have usually discussed previously with the Managing Director.'

(*v*) The fifth man had previously worked under the industrial scientist, being mainly concerned with the development of his designs for manufacture, but had recently moved into the position of technical sales manager. This made him for many purposes formally subordinate to the Sales Director and to the Commercial Manager of the firm. But he saw little of the first and avoided the second. He went to the head of the firm for 'sanctions, decisions, or instructions'.

Inevitably, where it was not clear who had the information or power to make decisions about delivery, programme priorities, sales effort, and other recurrent problems, the tendency was for each person concerned to go to the one person who unmistakably did have the power to decide. More important, in practice, was the fact that there were good reasons for not consulting others, because it would have the appearance of applying to someone in a better position to know, and therefore superior in status, or more powerful, or—which comes to the same thing—with better access to the head of the firm and a bigger share of his trust.

To consult others had the appearance of abdicating from one's own claims to highest status and most power in the department. At the same time, since, in all the ambiguous situations which arose, the closer one's association with the chief executive the more authority there was for actions, there was a constant temptation to involve him in any question relating to disputable fields of authority. This lay behind the 'by-passing' of certain executives of which they complained.

By-passing led directly to incidents which provoked a great deal of indignation. Entering the industrial scientist's office one morning, the interviewer was taken to the window. There was a large packing case, not yet closed up, waiting on a truck. It contained an equipment which 'everybody' had thought was to be dispatched the week following, but which the 'co-ordinating manager' had 'got the head of the firm to agree' should be dispatched the next day. 'Why, I don't know. There it is on the lorry now.' He led the way downstairs.

Outside, he pointed out instances of poor finishing, a wrongly coloured panel fitted, and so on. 'It just isn't good enough. People judge these things by the way it looks. . . .'

The same thing had happened over his classes, he said, on another occasion. 'They weren't really training classes—just to give background information to the younger ones on scientific matters which relate to what we're doing.' The arrangement was that half the time spent in classes should be counted as working time. This had been agreed by the head of the firm, and by other senior people concerned with staff and pay. The day before, the Production Manager, 'finding' that some of his people had left early, had gone to the Managing Director and had 'got him just to change the whole thing' and stop people going in the firm's time. 'Today', he said, 'I have to go round and tell these people that although I told them that they should come in the firm's time, they are not now able to, and I shall look a fool'.

Among the senior members of the firm, there were some who reacted against the competitive aspect of relationships with the Managing Director. They did not wish to appear to compete with their juniors in 'catching the boss's eye'. Lunchtime conversation in the absence of the head of the firm sometimes bore on this. On one occasion one member of the senior group, an informed and enthusiastic follower of football, spoke of the cup-tie to be replayed nearby that afternoon. The visiting team had 'only two players to watch. ―――― is one of the two dirtiest players in Scottish football. Of course, he's getting an old man now. You can see him getting in at the younger fellows with his elbows, edging them off, getting them rattled'. One of the others, exploding with laughter, said 'Is it football you're talking about?'; a remark which met with instant appreciation around the table.

A good deal of cynicism prevailed at this time about the policy of the Board and the prospects of the firm. One senior man kept a letter of resignation ready in his desk; another referred frequently to how well executives who had resigned in recent years had done for themselves. Some at a junior level had also withdrawn from the competition; one said at the end of an interview 'You never believe that the work you do will get you anywhere . . . there's little feeling that there is a real career depending on hard work and technical ability open to one here.'

Complementary to the attempts to capture authority and approval from the head of the firm by involving him in decisions was the tendency to displace responsibilities on to him. The 'knacky' welder who had proved so essential for some of the work done in the main shop for a new product (see p. 244) could have been replaced by a recruit or by further training for one of the other welders; but the problem had been 'posted higher up' and nothing done once the head of the firm had been informed; the executive function was regarded as discharged by merely awaiting the answer, even if it was getting overdue.

The conduct of each individual executive, including the head of the firm, may be regarded as a more emphatic version of the conduct which obtained formerly, in the days when it was quite compatible with the efficient operation of the concern as a whole. It was now incompatible with effectiveness. Nevertheless, they clung to the old form of the managerial system because it provided protection against the involvements the new order demanded of them. There was still only one man one had to bother with, only the activities embodied in instructions from him that devolved on one, only those decisions which were unchallengeable by anyone else that one made, only one man whose approval was necessary for recording achievement. One could use the system either to press one's claims for advancement or to protect oneself from heavier commitments and dubious undertakings. How little positive pressure there was for the liaison between executives, the absence of which was so frequently deplored, was demonstrated by the relative lack of success of a weekly progress meeting recently instituted. Attended by all executives concerned with one of the new products, the meetings were said to be concerned with 'very minor details'. Indeed, the meetings were said to be used to some extent as an excuse for not communicating outside them. And, it was also said, 'no real planning takes place' because everybody expected the head of the firm to initiate action without regard to the meetings.

The retention of the system of pair relationships in face of the requirements for collaborative action had its effect on the role of the head of the firm. Despite attempts to detach parts of his role, or to disperse them among a triumvirate, he was constantly drawn into making decisions not only about progress and deliveries, stores, sales, development and design details, but about the right to use the firm's car on this or that occasion. He spent two days supervising the erection of an exhibition stand. He felt obliged to do these things because there seemed no other way of settling the matter. Merely in order that work should proceed at all, he was compelled to act virtually as works manager, in addition to his role as Managing Director of the Company and to his share in the work of some three dozen subordinates. Through all this, he felt compelled to retain the position of directing agent he had formerly had, when each subordinate was a kind of machine-tool or instrument which had to be set by him every day.

What we are presented with here, then, is an extreme case of the principle by which each individual strove to shape the working organization according to his own conception of his needs and duties. For the individual himself, this could be done most economically by limiting 'the others' involved in each working occasion—each case for decision—to one. A form of management was therefore adhered to which consisted in the main of a system of relationships involving two people, with one always the head of the firm.

The context of the firm had become one in which decisions were so numerous, and so diffuse in character, that for it to survive effectively required each person to surrender a large part of his autonomy, and to exchange information and ally himself in decisions with extensive groups of people.

Yet, despite the perpetual irruption of faults, mishaps, and inconveniences for almost everybody, to 'get things done through one man' or to unload problems on to another retained all the attractions of simple direct action. And if the things were not done by one's superiors or subordinates, or if decisions were overruled, or if actions one relied on failed to occur, then, again, it was equally simple and direct to ascribe the fault to someone's personal defects. Despite the grievances and dissatisfactions, one avoided any deep commitment to the working organization. They were, in fact, the price paid for retaining one's detachment. Commitments grew around personal status and clique membership, not around the occupational role.

CHAPTER 12

Codes of Practice in Management Conduct

How a managing director deals with the facts of his own isolation and supremacy is related, it was said earlier, to the presence or absence of a system of beliefs about his capacities and his position (and their fitness for each other) common to him and his subordinates. The common system of beliefs existing in a concern, or in one section of a concern, is manifested most clearly in the way in which people treat each other. Each firm has its code of practice which defines the kind of conduct appropriate to managerial roles. The code is fixed—consciously or unconsciously, deliberately or by default—by the managing director. It is his most important contribution to the welfare of the organization. In the most successful concerns, the definition, maintenance, and control of admissible conduct is—in some form or other—one of his major preoccupations.

This was most clearly visible in those concerns with an organic system of management. The operation of an organic system of management hinges on effective communication. This is much more than a matter of providing, through the distribution of paper, for notification of events and decisions affecting functionally related persons and departments. It is also something more than providing for exchanges of information and opinion in meetings. What is essential is that nothing should inhibit individuals from applying to others for information and advice, or for additional effort. This in turn depends on the ability to suppress differences of status and of technical prestige on occasions of working interaction, and on the absence of barriers to communication founded on functional preserves, privilege, or personal reserve. In one or two firms, the existence of a way of behaving which facilitated this freedom of interaction was their most immediately obvious characteristic:

X: ' . . . Everybody is approachable by everybody else. It seems to me that it is almost a tradition here that that is so.'

Interviewer: 'And it is not particularly cultivated? . . . or is it cultivated?'

X: 'It is natural, I suppose, to some extent. People with long service in the organization just naturally do it, you see, and people coming in from outside, I suppose, just follow on. They can't do anything else, you see, because . . .'

Interviewer: 'You can't act stuffy among a crowd like that?'

X: 'No. No. You see, if the paint sprayer comes up with something and says, "Well, here you are chief", there's nothing much you can do about it, even if you want to. Which he does, you see; everybody calls everybody else "chief" here whoever they are. Except Mr. A. (a director)—he'll call everybody "governor".'

Interviewer: 'Well, that's a nice way of smoothing over status differences. If everybody is "chief", then there are no differences between people to bother about.'

X: 'Yes, that's right. Nobody can complain.'

A simple, obvious social device like this is of inestimable value. There are many occasions when differences of status, functional remoteness, different degrees of expertise make it difficult for questions to be asked or criticisms voiced. There are many other occasions when one is not sure whether the definition given by others to one's authority and function makes it possible for requests or decisions to be made; there are other occasions again when it is impossible to know whether another person will regard a task as his affair—whether a decision, an instruction, or a request for assistance will be resented as outside the self-imposed limits of his obligations to the firm. In all these situations, the ability to pitch what one says into a half-jocular style which explicitly rejects the pressure or sanction one could bring to bear is of enormous value. The social technique of doing this—the accepted formula used in this firm—was trivial. The fact that it could be successfully employed was all-important.

In another firm, the same results were produced by a more articulate, effortless manner. Equally, the facility and tact displayed in social interaction—as against the chumminess of the first concern—were the most immediately obvious characteristic. This is not to say that politeness and kindness are uncommon in industrial concerns—the writers have very good reason to know that they are not. But in this particular firm everyday occasions of meeting, conferences, informal discussions, and committee meetings were conducted according to sophisticated codes of manners. The amenities were observed. The widespread use of Christian names in addressing people up and down the status hierarchy was only the most superficial aspect of this social manner.

To some extent the manner derived from the location of the business in one of the most attractive suburban areas. People could be said to be living and working between the covers of the glossy magazines; the locale put a high premium on sociability and cultivated manners. But the dominant factor was, of course, that the code of conduct reflected that used by the managing director himself, who held strong beliefs about the respect due to other persons.

Such a manner is more than a lubricant. It is an essential part of the organization; in one sense, indeed, it is the organization, seen as a system of personal behaviour instead of a geometry of relationships. Change and growth brings with it new tasks, expansions of old tasks, and divided tasks. Some people are new, others promoted, others more or less stay where they were. The needs of this kind of situation are often read clearly enough for personal responsibilities and authority to be given as little specification from above as possible. The implications of this, however, are not only that any precise specification would be unrealistic or inapposite, not only that people could 'grow with the job', but also that the limits of defeasible action by an individual or of defeasible demands upon him are left much broader than is customary. By defeasible, in this connexion, is meant 'possible to resist or reject on the grounds that such action or requests to others are outside the contractual obligations of work in that position in the concern'. So, in order not to draw invidious attention to action, or requests, or instructions, or even calls for extra effort, which might lie outside what is normally expected of people in such jobs, there is a strong pressure for every move to be conducted in terms of friendly intimacy. The code of practice in conduct has, however, a latent function, which is of more direct importance. It is an indicator of the range of factors in a situation which may be openly admitted to exist.

Any discussion, any conflict, any decision is carried through because

of a number of considerations, of which some are expressible and regarded as having force, others are inexpressible and disregarded, in
overt communication. Most people will have experience of how technical, personal, and commercial considerations affecting an issue may be
ruled out by a restrictive style, by adherence to 'hard-headed' codes of
conduct, or—more often—by banter and irony used as a barrier to the
entry of intellectually or emotionally difficult notions into discussion.

An excerpt from a recording of a foremen's meeting may demonstrate at a fairly elementary level the conception we are trying to
present:

Chairman: 'Any other points?' (*Long pause.*)

Foreman A: 'There's a problem in my department. I've got two
turners; they're good men but they make no bonus—
but they'll not scrap any jobs. They've been in about
four or five months now and I don't think they've
scrapped a job between them. But they're not looking
at their times.'

Chairman: 'Then they're not good men, are they? . . . A good man
is a man who makes a bonus and makes no scrap. Now
these men make no scrap; but they make no bonus.
Therefore they're only half good.'

Foreman A: 'Well—what do we do?'

Chairman: 'Well, we pay them by results. That's our accepted
system here. . . . Suppose we pay these men a large
bonus for not working hard: what happens to the other
people?'

Foreman A: 'You see, the ratefixer's been agitating at me to see if I
could get these men—to see what was the matter with
them that they're not making bonus.'

Chairman: 'What is the matter with them?'

Foreman A: 'Well, it appears to me, they're just made that way.'

Chairman: 'Do they want to make any bonus?'

Foreman A: 'Well, it doesn't seem to worry them—they never do,
anyway.'

Chairman: 'Why cross that bridge before we come to it, then?'

Dpt. Mgr. 'Isn't it quite clear that if these men are consistently not
A: earning bonus, then they're not much use to us anyway?'

Foreman B: 'Of course, these men are average men, and that's what bonus is fixed on.'

Foreman A: 'They're good, these men.'

Dpt. Mgr. A: 'If they're average men then they should be earning average bonus.

(Chorus of dissent—confused argument about quality of the men's work.)

Dpt. Mgr. A: 'From the job point of view, they're first rate, that's the point. But they appear to be individuals with no great desire to earn more pennies. They also enjoy a fairly leisured existence and do high class work.'

Foreman C: 'I shouldn't like to say that the man who makes most bonus is the best man. Speed isn't always efficient—it's effective effort. . . .'

Dpt. Mgr. A: 'Surely the question really is, do we have the class of work suited to men of this description?'

Foreman C: 'Evidently, according to the amount of scrap that's being produced around the factory, it's advisable to have a few of these men in the shop.'

Dpt. Mgr. B: 'They fill their part, really, I think.

Chairman: 'But people who make bonus don't make scrap.'

Foreman C: 'It is the people who make bonus that make scrap—not these people who are slow but efficient.'

Chairman: 'Are you talking about your department?'

Foreman C: 'No. It's general through the factory.' *(Chorus of assent.)*

Foreman D: 'You get this type of chap in every department—they aren't worried about money so much as turning out a good job. They've got pride in their work.'

Chairman: 'They're not complaining they're not making any money, are they?'

Foreman A: 'No, but this ratefixer, he puts a time on the job and they're taking far in excess of the time.'

Chairman: 'Well then, they're below average. And they're quite happy about not making bonus, and we're quite happy —at the moment—to have our machines run at a low utilization rate. If the time comes when we're short of machines, then we'll have to consider doing something about it.'

Foreman A: 'That's all I wanted to know—the general policy.'

(This was followed by a long pause with the feeling pretty apparent on all sides that the subject had not been disposed of. The discussion was then resumed, mainly between the departmental managers, but the same position was maintained by the Chairman until the end.)

In this episode, the two sides of the argument in fact represented two different spans of consideration. The wider span was invoked by the foremen, who wanted to discuss the adequacy of the rate-fixing and bonus system used in the factory as a means of measuring and recompensing the total effective contribution of operatives' work. By setting the limits of admissible considerations much more narrowly, the chairman was able to exclude this whole question from discussion. He did so by employing a style of conduct in which banter covered, fairly thinly, a deliberate invocation of his authority to 'speak for the firm' (i.e. the management), an authority which was supported fairly clearly by the belief of the foremen. His employment of the word 'we' at a critical point of the discussion was of the first importance.

We need not decide whether he was right to do so or not. The point of the quotation, it will be remembered, is to demonstrate the way in which the bounds of admissible considerations in any decision-making situation may be—as they usually are—set by superiors. The span of such considerations for normal use throughout the concern tends to be set by the managing director. The effect is to specify the terms in which questions can be discussed and also the factors of which decisions must take account. The narrower the limits, the easier it is to deal with problems and to reach decisions. This is the whole point of formalism, of rules and set procedures, of mechanistic organization—that it makes managerial functions much easier and simpler. It 'programmes' managerial decisions.

In the circumstances prevailing in firms such as those which were the subjects of these studies, such simplicity is usually deceptive; or rather, it exacerbates organizational and personal problems by refusing to acknowledge their reality. No firm has found a style of conduct ideally adapted to the requirements of its situation. This may account for the impression derived from both Scottish and English studies that anxiety and stress was most evident in those firms which were expanding most rapidly and were technically and commercially most progressive. It was even held by some informants that an abnormally high level of emotional stress was a necessary element of the success of these firms.

Doubtless, the operation of an organic form of management system tends to foster the insecurities which give rise to such stresses. Indeed, negative evidence of the use of organic procedures was available in many firms in the complaint, by someone or other, that jobs and responsibilities, or the line of authority, were never clearly enough defined. And it was always possible for people to seem to reduce their insecurity relative to others by trying to render others even more insecure, and so direct their energies to weakening the positions of other people around them by attacking their self-confidence and lowering their prestige. This kind of response tended to be mingled with the other legitimate, desired, response, so that the firms which were effervescent and exciting places to work in could also be harmful and exhausting. What did help to reduce strain was the admission of open and realistic discussion of the psychological difficulties of a situation. The doorkeeper who controls the admission of such considerations, as of other kinds, is the managing director.

COUNTER-SYSTEMS

Every firm is a community, with its own particular flavour, its own social structure, its own style of conduct. Newcomers are very conscious of this quality of uniqueness. Indeed, they have to be, since they have to learn the culture, and until they do, until it is other places which begin to have a disconcertingly unfamiliar smell, they have neither been accepted nor accepted their position.

The very close cohesion, obvious in one firm, and clearly expressed in the 'happy family' manner of dealing with each other to which we have referred, made it particularly difficult for it to absorb newcomers. This does not mean, though, that new recruits were cold-shouldered, or that the 'old guard' was exclusive. As we saw it, in connexion with two senior managers introduced in the previous year or two, the difficulties arose out of the perception by the newcomers of the particular style of familiar intimacy prevailing among the established staff—a style which allowed all kinds of criticism and appeals for co-operation to be voiced without strain. It was inevitable for them to feel 'odd man out in many working situations which involved them. It was almost as inevitable that, in order to fit in, they adopted as quickly as possible a resemblance of the manner they had observed used to much effect, or a substitute for it. This turned out in practice to be either a false and nervous bonhomie or a carefully articulated courtesy. Either manner grated on the older members of the staff, and there was an admitted tendency to fence the newcomer off so far as the working organization was concerned. This was a simple matter of making use of the open communication system to bypass him and thus make him semi-redundant. This had the effect of engendering doubts in his superiors—and eventually in himself—about his value to the firm. The final stage was an obvious diffidence about making the special contribution to the firm for which he had been recruited. Thus he justified the efforts to isolate him, and led the directors to feel that they had somehow been deceived.

The code of practice for conduct was described in the previous section as both revealing and acting as the medium for specifying the limits of defeasible action, whether they are elastic or fixed, and for intimating what span of all the rational, technical, cultural, and emotional elements of a situation or problem may be counted as admissible. The existence, the character, and the functions of such a code of practice are most easily realized by observing the impact of newcomers on it, but it is also made apparent by the contrast of a counter-style developed by any powerful political minority movement within the firm.

The essentially sympathetic and articulate style characteristic of one firm was explicitly rejected by a few individuals and tough-minded insensitivity, or the brash 'positive personality' of the traditional salesman, substituted for it. This not only damaged personal relationships, but actually hampered the essential working of the managerial system. The incompatibility which it bespeaks was responsible, for example, for a situation originating in the appointment of a publicity manager by two directors with only 'moderate participation' from the Sales Director, a leading exponent of the counter-system; this gave rise to a campaign against publicity, led by sales, which had the added attraction of a chance to make common cause with technical and other groups who would usually be felt to be in opposition to sales.

Situations of this kind may grow to the point of there being an orthodox and a nonconformist culture, code of conduct, or system of 'ideas about the right way and the wrong way to go about things'. Such differences provide the setting and the ideologies for political dissensions. Indeed, in this firm it was the main arena of political conflict. The views of the managing director concerning decision-making by consultation, his dislike of a command hierarchy, his preference for an organic system, his conception of the commercial purposes and sales policy appropriate to the firm, his resistance to the growth of privileges and perquisites, were regarded by some members of the staff as utopian or unrealistic, and at odds with what actually happens.

What was really at stake in these and other matters—promotion, selection, the institution of managerial techniques, the direction and nature of sales effort—was the simple political issue of the amount of control, of 'say' in those matters of common concern to the whole company, which might be vested in the managing director as against his colleagues. Control, or 'say', is obviously not 'unchallenged authority'—for one of the main positions defended by the managing director was the practice of promoting effective action through decisions arrived at by the persons who were involved as agents. Issues over control, or 'say', had to do rather with the extent to which others felt they could themselves enjoy 'unchallenged authority' within specific spheres of their own, with the extent to which they would allow themselves to become convinced of the rightness of views advanced by the managing director, with the extent to which his approval was regarded as the one essential requisite for any course of action to be decided upon.

There is in management, as in any other working group, a perpetual pressure on individuals to adjust their roles, and the manner in which they perform them, to accord with the changing exigencies of material circumstances and with the pressures exerted by other members of the group. Everybody is 'becoming a different person' in this way all the time. These changes are often irksome, sometimes depressive; always, however, they will be regarded as invasions or expansions of the individual's field of established competence. The effort made by a managing director to persuade others of the rightness of the system which permits these things to happen, the effort to resist him, or to demonstrate that his views are wrong, underlie many conflicts. Ideological victory, after all, carries with it a measure of political control as nearly absolute as one can hope for.

CONDUCT AND SOCIAL CONTROL

The code of conduct, the considerations admitted to discussion and decision-making, the system of beliefs about the situation of the firm and the capacities of its members, the nature of relationships with superiors, equals, and subordinates, and the definition given to authority, may each be conceived as an aspect of the others. All have to do with the conduct of people towards each other, so that their interrelation is no more than self-evident, in one way. What is being suggested here, however, is not merely that they are isolable features of human behaviour, but that they are different sets of words to describe the same facts of observation. As they have been set down above, the differences are arranged in an order of decreasing precision and meaningfulness.

The limits of the significance to a style of conduct, and all it comprehends, are set by the institutional environment within which people act; i.e. the market situation, the organizational structure, the total disposable resources of the firm, the existing organizational system, the political and status structure, the directional policy. All these are subject to change and alteration. Management is concerned largely with anticipating change and making alteration. What we have subsumed under the general notion of code of conduct may be regarded as the dynamics of alteration within the limits of the institutional environment.

In contesting the amount of say one person is to have over his own activities and those of others, decisions are made in accordance with the considerations admitted as relevant. The code of conduct accepted as dominant in the situation defines the span of considerations which are in fact admitted. This is also true, as we have tried to show, for the innumerable executive occasions when decisions have to be arrived at by means of an interpretation of directional policy. It is more than ever true of decisions about the directional policy itself, since these are typically occasions of planning action and arrangements for the future. Many of the most important elements which will determine the ultimate success or failure of that action are either unknown or are only assessable in crude terms, or are impossible to rank in order of importance. It is when decisions have to be made on a knowledge of the facts inevitably insufficient for assurance that the question of the dominance of this or that set of considerations becomes of paramount importance.

CONFLICT AND THE OVERRIDING TASK

The possibilities of controlling the direction of decisions through the imposition of a dominant style of conduct are, as we have said, circumscribed by the actualities of the situation. But they are so circumscribed only in so far as the actualities are perceived by the people concerned.

There is, the managing director of one concern remarked, an inverse ratio between the intensity of internal conflict and the seriousness of the problems and tasks facing the organization. This, he said, had been amply demonstrated nationally during the war; it was true also of ordinary families in the emergencies caused by illness and other threats. But 'serious' problems are those which 'force' themselves on the awareness of those whom they affect. He was convinced that one of the most important jobs of management was constantly to inform the members of the concern of the situation confronting them as a community—of the circumstances of the market, of technical developments, of conditions in industry at large, of competitive pressures, of elements in national and international trading conditions, and of the changes in them.

This book has, in fact, dealt with an array of internal manifestations of the external tasks and problems, and of changes in their disposition, which affect the existence of the concern as a whole. There is an obligation on management not only to interpret the external situation to the members of the concern, but to present the internal problems for what they truly are: the outcome of the stresses and changes in that situation —in markets, technical requirements, the structure of society itself.

References

1. BURNS, TOM. 'The Reference of Conduct in Small Groups; Cliques and Cabals in Occupational Milieux.' *Human Relations*, 8 (1955), pp. 467–86.
2. ARGYRIS, C. *Executive Leadership*. New York: Harper, 1953.
3. MARCUSE, H. *Eros and Civilization*. London: Routledge, 1956.
4. OGBURN, W. F. *Social Change*. New York: Viking Press, 1922.
5. BOULDING, K. E. *The Organizational Revolution*. New York: Harper, 1953.
6. MARX, K. 'Letter to P. V. Ennenkov', 1846. In: Karl Marx and Frederick Engels, *Selected Works*, Vol. II. London: Lawrence and Wishart, pp. 401–2.
7. DURKHEIM, E. *De la Division du travail social*. 1893. Trans.: *On the Division of Labour in Society*, by G. Simpson. Glencoe, Ill.: Free Press, 1954.
8. TÖNNIES, F.: *Gemeinschaft und Gesellschaft*, Leipzig, 1887. Trans.: *Fundamental Concepts of Sociology*, by C. P. Loomis. New York: American Book Company.
9. JEWKES, J. 'How much Science.' Presidential Address to British Association, Economic Section, 1959. *Economic Journal*, No. 277. March 1960, **70**, p. 12.
10. WHITEHEAD, A. N. *Science and the Modern World*. London: Cambridge Univ. Press 1926. (7th Impression, 1933, p. 120.)
11. BRIGHT, A. A. *The Electric Lamp Industry: Technological Change and Economic Development from 1880 to 1947*. London: Macmillan, 1949.
12. MACLAURIN, W. R. *Invention and Innovation in the Radio Industry*. New York: Macmillan, 1949.
13. CLOW, A., and CLOW, N. *The Chemical Revolution*. London: Batchworth, 1952 (pp. 593–4).
14. SMILES, S. *Life of Boulton and Watt*. London: Murray, 1865 (p. 367).
15. *British Museum Catalogue of Printed Books*, Vol. 36 (Periodicals: Enlarged Edn.), 1899.
16. MAYO, E. *The Social Problems of an Industrial Civilization*. London: Routledge, 1949 (p. 32).
17. FERRANTI, G. Z. DE, & INCE, R. *The Life and Letters of Sebastian Ziani de Ferranti*. London: Williams and Norgate, 1934 (pp. 51–2).
18. SNOW, C. P. *The Two Cultures and the Scientific Revolution*. London: Cambridge University Press, 1959.
19. BERNAL, J. D. *Science and Industry in the Nineteenth Century*. London: Routledge, 1954 (pp. 63–4).
20. CARDWELL, D. S. L. *The Organization of Science in England*. London: Heinemann, 1957.
21. Department of Scientific and Industrial Research: *Estimates of Resources devoted to Scientific and Engineering Research and Development in British Manufacturing Industry*, 1955. H.M.S.O., 1958.

22. Treasury: *Civil Estimates*, 1918–51.
23. BERLE, A. A., JR., and MEANS, G. C. *The Modern Corporation and Private Property* London: Macmillan, 1932.
24. FRIEDMANN, W. *Legal Theory*. London: Stevens, 1949 (p. 244).
25. FRIEDMANN, W. *Law and Social Change in Contemporary Britain*. London: Stevens, 1951 (p. 15).
26. KEIRSTEAD, B. S. *The Theory of Economic Change*. London: Macmillan, 1948 (p. 254).
27. WHITTLE, H. *Jet*. London: Muller, 1954.
28. SCOTT, J. D., and HUGHES, R.: *The Administration of War Production*. H.M.S.O., 1955 (p. 378).
29. CARTER, C. F., MEREDITH, G. P., and SHACKLE, G. L. S. (eds.). *Uncertainty and Business Decisions*. Liverpool University Press, 1954. (See Frontispiece illustration of the Shackle model.)
30. SIMON, H. A. *Administrative Behaviour*. New York: Macmillan, 1957.
31. BLACK, D. *Theory of Committees and Elections*, 1958.
32. ROETHLISBERGER, F. V. AND DICKSON, W. J. *Management and the Worker*. Harvard Univ. Press, 1939. Pt. IV, 'Social Organization of Employees'.
33. TAYLOR, F. W. *The Principles of Scientific Management*, 1911. (Reprinted in *Scientific Management*. New York: Harper, 1947.)
34. BURNS, TOM. 'The Forms of Conduct', *American Journal of Sociology*, **64**, 1958, p. 148.
35. GOULDNER, A. W. 'Organizational Analysis.' In R. K. Merton, L. Broom and L. S. Cottrell (eds.), *Sociology To-day*. New York: Basic Books, 1958.
36. SELZNICK, P. *T.V.A. and the Grass Roots*, University of California Press, 1948.
37. WHYTE, W. F. *Street Corner Society*, University of Chicago Press, 1943.
38. HOMANS, G. F. *The Human Group*. London: Routledge, 1951 (p. 169–171).
39. BARNARD, C. I. 'The Functions and Pathology of Status Systems in Formal Organizations'. In *Organization and Management*. Harvard Univ. Press, 1946 (p. 243).
40. BURNS, TOM. 'The Directions of Activity and Communication in a Departmental Executive Group,' *Human Relations*, **7** (1954), pp. 73–97.
41. SELZNICK, P. *Leadership in Administration*. Evanston, Ill.: Row, Peterson, 1957 (pp. 135–6).
42. FOLLETT, MARY P. *Dynamic Administration* (ed. Metcalfe & Urwick). Management Publications Trust, 1941.
43. BENDIX, R. *Work and Authority in Industry*. London: Chapman & Hall, 1956.
44. CLAPHAM, J. H. *Economic History of Modern Britain*, Vol. II: Free Trade and Steel 1850–86. London: Cambridge Univ. Press, 1932.
45. BARNARD, C. I. *The Functions of the Executive*. Harvard Univ. Press, 1938.
46. BURIN, F. G. 'Bureaucracy and National Socialism.' In Merton, R. K., *et al.* (ed.) *Reader in Bureaucracy*. Glencoe, Ill.: Free Press, 1952 (pp. 33–47).
47. JÜNGER, E. *Der Arbeiter*, Hanseatische Verlag, 1932.
48. WEBER, M. *The Theory of Social and Economic Organization* (tr. Henderson and Parsons). W. Hodge, 1947 (pp. 329–34).
49. FAYOL, H. *Industrial and General Management*. (Trans.) Pitman, 1948.
50. NORMAN, W. H. *Administrative Action, the Techniques of Organization and Management*. Pitman (9th U.S. edn.), 1958.
51. BRECH, E. F. L. *Principles and Practice of Management*. London: Longmans, 1953 (see p. 25).
52. MILLER, D. C., and FORM, W. H. *Industrial Sociology*. New York: Harper, 1951.

53. ROETHLISBERGER, F. V. *Management and Morale.* Harvard University Press, 1941.
54. MOORE, W. G. *Industrial Relations and the Social Order.* New York: Macmillan. 1947.
55. WALDO, D. *Perspectives on Administration.* Univ. of Alabama Press, 1956.
56. GOULDNER, A. W. *Patterns of Industrial Bureaucracy.* London: Routledge, 1956.
57. EISENSTADT, S. N. 'Bureaucracy and Bureaucratisation'. In: *Current Sociology,* 7, No. 2 (Blackwell) (1958), p. 106.
58. LEACH, E. R. *Political Systems of Highland Burma.* London: Bell, 1954 (pp. 284–5).
59. ALLPORT, F. H. *Theories of Perception and the Concept of Structure.* New York: Wiley, 1955.
60. DEUTSCH, MORTON. 'An Experimental Study of the Effects of Co-operation and Competition upon Group Processes'. *Human Relations,* 2 (1949), pp. 199–232.
61. EDWARDS, WARD. 'Probability-Preferences in Gambling'. *American Journal of Psychology,* 66 (1953), pp. 349–64.
62. CHRISTIE, L. S., LUCE, R. DUNCAN, and MACY, J., JR. *Communication and Learning in Task-Oriented Groups.* Research Laboratory of Electronics, Massachusetts Inst. of Technology Tech. Rept. No. 231, 1952.
63. BALES, R. F. *Interaction Process Analysis.* New York, 1950.
64. BAVELAS, ALEX. 'Communication Patterns in Problem Solving Groups.' In: H. von Foerster (ed.) *Cybernetics, Transactions of the Eighth Conference 1951,* Josiah Macy Jr. Foundation, 1952.
65. CARTER, C. F., MEREDITH, G. P., and SHACKLE, G. L. S. (eds.). *Uncertainty and Business Decisions.* Liverpool Univ. Press (p. vii.).
66. SHACKLE, G. L. S. *Expectation in Economics.* Cambridge Univ. Press, 1949.
67. SHACKLE, G. L. S. *Uncertainty in Economics.* Cambridge Univ. Press, 1955.
68. SHACKLE, G. L. S. 'Expectation and Liquidity.' In: Bowman (ed.) *Expectations, Uncertainty and Business Behavior.* New York: Social Science Research Council, 1958.
69. GEORGESCU-ROEGEN, N. 'The Nature of Expectation and Uncertainty.' In: Mary J. Bowman (ed.) *Expectations, Uncertainty and Business Behavior.* New York: Social Sciences Research Council, 1958 (p. 14).
70. GALLIE, W. B. 'Uncertainty as a Philosophical Problem' in Carter, Meredith and Shackle (eds.), *op. cit.*[65] (p. 3).
71. SHACKLE, G. L. S. Final Comment in C. F. Carter, G. P. Meredith and G. L. S. Shackle (eds.) *Uncertainty and Business Decisions,* Liverpool Univ. Press, 1954 (p. 100).
72. TOULMIN, S. *The Uses of Argument,* Cambridge University Press, 1958 (Chap. 2).
73. CARTER, C. F. 'A Revised Theory of Expectations' in C. F. Carter, G. P. Meredith and G. L. S. Shackle, *op. cit.*[65] p. 54.
74. SIMON, H. A. 'The Role of Expectations in an Adaptive or Behavioristic Model'. In Bowman (ed.) *Expectations, Uncertainty and Business Behavior.* New York: Social Science Research Council, 1958.
75. CHERRY, C. *On Human Communication.* New York: Wiley, 1957.
76. ECCLES, C. 'The Physiology of Imagination.' *Scientific American,* 99, No. 3, Sept., 1958.
77. MEREDITH, G. P. 'A Revision of Spearman's Neogenetic Principles.' In: *Proc. Aristotelian Society,* 49 (N.S.), 1949.
78. MEREDITH, G.P. 'The Surprise Function and the Epistemic Theory of Expectations. In: Bowman (ed., *Expectations, Uncertainty and Business Behavior.* New York: Social Science Research Council, 1958.
79. HENRY, W. G. Comment on G. P. Merediths' paper in Bowman (ed.), *Expectations, Uncertainty, and Business Behavior.* New York: Social Science Research Council, 1958, p. 84.

80. BURNS, TOM. 'The Idea of Structure in Sociology.' *Human Relations*, 9, 1958, 220.
81. PARSONS, T. 'The Superego and the Theory of Social Systems.' In: Parsons, T., Bales, R. F., and Shils, E. A., *Working Papers in the Theory of Action*. Glencoe, Ill.: Free Press, 1952.
82. KLEIN, J. *The Study of Groups*. London: Routledge, 1956 (chap. 2).
83. HAIRE, M. *Psychology in Management*. New York: McGraw Hill, 1956 (p. 54).
84. SHEPARD, H. A. 'Superiors and Subordinates in Research' (Paper 12 of the *Symposium on the Direction of Research Establishments*). H.M.S.O.: Department of Scientific and Industrial Research.
85. ARENSBERG, C. M., & MACGREGOR, D. 'Determination of Morale in an Industrial Company.' *Applied Anthropology*, I, 1942, pp. 12–34.
86. JEWKES, J., SAWERS, D., and STILLERMAN, R. *The Sources of Invention*. London: Macmillan, 1957 (pp. 94–5).
87. BURNS, TOM. 'Management in Action.' *Operational Research Quarterly*, 8 (1957), pp. 45–60.
88. RICE, A. K. *Productivity and Social Organization: The Ahmedabad Experiment*. London: Tavistock Publications, 1958.
89. SIMMEL, G. *The Sociology of Georg Simmel* (ed. K. Wolf). Glencoe, Ill.: Free Press, 1950 (pp. 287–300).

Index